Essays in Biochemistry

volume 29 1995

Essays in Biochemistry

Edited by D.K. Apps and K.F. Tipton

PORTLAND PRESS

Essays in Biochemistry is published by Portland Press Ltd on behalf of the Biochemical Society

Portland Press Ltd
59 Portland Place
London W1N 3AJ
U.K

British Library Cataloguing-in-Publication Data
A catalogue record for this book is available from the British Library

ISBN 1-85578-017-8
ISSN 0071-1365

Typeset by Portland Press Ltd and printed in Great Britain by Henry Ling Ltd, Dorchester

Contents

1 **Bacterial DD-transpeptidases and penicillin**
   Marc Jamin, Jean-Marc Wilkin and Jean-Marie Frère

2 **Sphingolipid activator proteins**
   Kunihiko Suzuki

3 Oxygen toxicity, free radicals and antioxidants in human disease: biochemical implications in atherosclerosis and the problems of premature neonates
Catherine A. Rice-Evans and Vimala Gopinathan

4 Reconstructed human skin: transplant, graft or biological dressing?
Edward J. Wood and Ian R. Harris

5 Opsin genes
B. Edward H. Maden

6 **The roles of molecular chaperones** *in vivo*
Peter A. Lund

7 **Molecular chaperones: physical and mechanistic properties**
Steven G. Burston and Anthony R. Clarke

8 **Affinity precipitation: a novel approach to protein purification**
Jane A. Irwin and Keith F. Tipton

9 **Molecular pathology of prion diseases**
Corinne Smith and John Collinge

10 Ribozymes
Helen A. James and Philip C. Turner

11 Protein stability at high temperatures
D.A. Cowan

The Authors

Jean-Marie Frère graduated from the University of Liège in 1964 with a degree in Chemistry and obtained a Ph.D. in Biochemistry from the University of Montréal in 1969. After holding a postdoctoral position for one year at the Massachusetts Institute of Technology, he returned to Liège in 1970 and was appointed Professor of Enzymology in 1979. Since 1971, his research interests have mainly centred on the enzymes that interact with β-lactam antibiotics: penicillin-sensitive DD-peptidases and β-lactamases. **Marc Jamin** graduated in 1988 in Biochemistry from the University of Liège. He obtained a Ph.D. in Biochemistry in 1993 under the supervision of J. M. Frère. He currently holds a postdoctoral position in the Department of Biochemistry at Liège as a Chargé de Recherches of the Belgian National Research Foundation. **Jean-Marc Wilkin** obtained his B.Sc. in Biochemistry from the University of Liège in 1989 and his Ph.D. (under the supervision of J.M. Frère) in 1993. He is currently working as a postdoctoral fellow at the Research School of Biological Sciences, The Australian National University, Canberra.

Kunihiko Suzuki, M.D., graduated in History and Philosophy of Science from Tokyo University, Japan, in 1955 and obtained his M.D. from the Faculty of Medicine in 1959. During the 1960s he held academic positions at Albert Einstein College of Medicine, New York, and the University of Pennsylvania School of Medicine. In 1972 he was appointed Professor of Neurology and Neuroscience at the Albert Einstein College of Medicine. He has held positions as Director of the Brain and Development Research Center, and Professor of Neurology and Psychiatry at the University of North Carolina School of Medicine, since 1986. His scientific interests have included brain lipids and their metabolism, and genetic neurological disorders, particularly those involving lysosomal hydrolases. He has actively participated in the evolution of the field through the phases of analytical biochemistry, enzymology and molecular biology.

Catherine Rice-Evans is Professor of Biochemistry at UMDS, Guy's Hospital, and Director of Free Radical Research. She obtained her B.Sc. degree in Chemistry at the University of London and her Ph.D. undertaking research at the National Institute for Medical Research. Her major research interests are in the involvement of free radicals in the pathogenesis of disease and the role of antioxidants in the maintenance of health and disease prevention. She is currently President of the Society for Free Radical Research (Europe) and a member of UNESCO Global Network for Molecular and Cell Biology. **Vimala Gopinathan** is Senior Registrar in Paediatrics at St Thomas's Hospital with a special interest in neonates. She became a member of the Royal College of

Physicians of London in 1989. Her major research interests are in plasma antioxidant status in preterm infants.

Ed Wood has a D.Phil. from the University of Oxford and has worked at the University of Leeds since 1971, where he is now Reader and Head of the Department of Biochemistry and Molecular Biology. His research interests are in human skin and its diseases, and in wound healing. He is also Editor of *Biochemical Education* and Chairman of the Professional and Education Committee of the Biochemical Society. **Ian Harris** graduated in Biochemistry at the University of York, 1990, and received a Ph.D. from the Department of Biochemistry and Molecular Biology at the University of Leeds in 1994. His interests in using keratinocyte sheets as a treatment for patients with leg ulcers have continued his work at the Yorkshire Regional Tissue Bank.

Edward (Ted) Maden is Johnston Professor of Biochemistry at the University of Liverpool, where he has recently introduced an Honours Module course on Specific Eukaryotic Genes. This course complements a standard course on gene structure and expression, by showing for a hand-picked selection of eukaryotic genes how the cloning of these genes arose from the contemporary state of the art, and what has been learnt from their detailed and continuing analysis. Genes chosen for this kind of in-depth treatment include ribosomal RNA genes, on which Ted Maden has carried out most of his research, globin genes, in which many fundamentals of eukaryotic genes were first discovered, and opsin genes, which underlie our visual perception of the external world, as well as several other exemplary gene systems. The seeds for Ted Maden's interest in opsins and visual perception were sown in his final year at Cambridge University where, as a Physiology undergraduate, he attended a series of lectures on vision by the late W.A.H. Rushton.

Peter Lund is a lecturer in the Microbial Genetics and Cell Biology Research Group, in the School of Biological Sciences at the University of Birmingham. His current research interests encompass several areas of molecular chaperone and heat shock protein biology, in particular the biological role of the hsp60 proteins and their potential medical and biotechnological importance. He has published work in several different research areas, including protein secretion, gene targeting, and regulation of gene expression.

Anthony R. Clarke obtained his B.Sc. from the University of Sheffield in 1980, and his Ph.D. from the University of Bristol in 1983. He is currently a lecturer and Lister Institute Fellow at the University of Bristol. He has published extensively in the field of enzyme catalysis and protein engineering and, more recently, on protein folding and the mechanism of chaperonins. **Steven G. Burston** obtained a B.Sc. in 1989 and a Ph.D. in 1992 from the University of Bristol, where he is currently a postdoctoral associate in the Biochemistry Department. His initial work centred on identifying equilibrium and kinetic intermediates during spontaneous protein folding, and he has recently entered the field of molecular chaperones.

Jane Irwin obtained her degree in Biochemistry from the University of Dublin in 1987 and her Ph.D. in 1994. She is currently employed in the Department of Biochemistry in a project concerned with the diagnosis of Alzheimer's disease. **Keith Tipton** graduated with a degree in Biochemistry from the University of St Andrews in 1962, and obtained his Ph.D. from the University of Cambridge in 1966. He is currently Professor of Biochemistry at Trinity College, Dublin.

Corinne Smith is a postdoctoral scientist working in the group of Dr Tony Clarke in the Department of Biochemistry at the University of Bristol. She is a visiting worker with the Prion Disease Group. **John Collinge** is a Wellcome Senior Research Fellow in the Clinical Sciences and leads the Prion Disease Group at the Department of Biochemistry and Molecular Genetics at St Mary's Hospital Medical School, Imperial College, London.

Phil Turner is a Senior Lecturer in the Department of Biochemistry, University of Liverpool, where his research group is involved in studying various aspects of gene expression in eukaryotes. These include studies of transcription factors, in both animal viruses and their hosts, the function of the U7 small nuclear RNP particle in processing histone pre-mRNA and the use of ribozymes to inhibit gene expression. He is a graduate of the Biochemistry Department at Leeds University and obtained his Ph.D. in the Department of Molecular Biology, University of Edinburgh, in 1978. Before being appointed to his present position, he was a member of the Department of Biological Sciences, University of Warwick, where he carried out postdoctoral work on histone gene expression in the Developmental Biology Group. **Helen James** graduated in 1990 from the University of Wales, Cardiff, with a B.Sc. Joint Honours in Biochemistry and Chemistry. She went on to study ribozymes and modified snRNAs for a Ph.D. at Liverpool University, in the laboratory of Dr Phil Turner, and is currently a postdoctoral research associate in the School of Biological Sciences, University of East Anglia. Her research involves the design and optimization of ribozymes against mRNAs associated with chronic myeloid leukaemia with a view to a potential therapy.

Don Cowan graduated with a Ph.D. from the University of Waikato in Hamilton, New Zealand, in 1980. He stayed with the newly formed Thermophile Research Unit, then funded by Shell Ventures, for a further 5 years, working on the enzymology of a number of thermophilic and hyperthermophilic Bacteria and Archaea. In 1985 he moved to the U.K. to take up a lectureship in the Department of Biochemistry at University College, London, where he currently holds a Senior Lectureship, supervises a research group (still working on thermophiles), administers a B.Sc. degree in Biotechnology, runs a small biotechnology company and seldom suffers from boredom.

Abbreviations

A_2pm	diaminopimelic acid
Ac_2KAA	N^α, N^ε-diacetyl-L-lysyl-D-alanyl-D-alanine
AcKAA	N^α-acetyl-L-lysyl-D-alanyl-D-alanine
ADA	adenosine deaminase
ADH	horse liver alcohol dehydrogenase
ADRP	autosomal dominant retinitis pigmentosa
apo E	apolipoprotein E
ASVB	Avocado sun-blotch virus
a.t.r.-f.t.i.r.	attenuated total reflection Fourier transform infrared spectroscopy
BH	protonated base
bis(NAD^+)	N_2, N_2'-(adipodihydrazide)-bis(N^6-carbonylmethyl)NAD^+
BPD	bronchopulmonary dysplasia
BSE	bovine spongiform encephalopathy
CAT	chloramphenicol acetyltransferase
c.d.	circular dichroism
CJD	Creutzfeld–Jakob disease
CLD	chronic lung disease
cpn	chaperonin
DD-peptidase	D-alanyl-D-alanine peptidase
DMSO	dimethylsulphoxide
ECM	extracellular matrix
EGF	epidermal growth factor
ER	endoplasmic reticulum
GDH	glutamate dehydrogenase
GPI	glycosylphosphatidylinositol
GSS	Gerstmann–Sträussler syndrome
HDV	hepatitis delta virus
h.p.l.c.	high-performance liquid chromatography
H_2O_2	hydrogen peroxide
HIV	human immunodeficiency virus
HLA	human leukocyte antigen
HMM	high molecular mass
hsp	heat-shock protein
IgG	immunoglobulin G
IVH	intraventricular haemorrhage
IVS	intervening sequence

LCR	ligase chain reaction
LDH	lactate dehydrogenase
LDL	low-density lipoprotein
LMM	low molecular mass
MHC	major histocompatibility complex
n.m.r.	nuclear magnetic resonance
NEC	necrotizing enterocolitis
NIPAM	*N*-isopropyl acrylamide
NO	nitric oxide
nt	nucleotide
$O_2^{-\cdot}$	superoxide radical
PBP	penicillin-binding protein
PCR	polymerase chain reaction
PEG	polyethylene glycol
PFK	phosphofructokinase
PrP	prion protein
RNase P	ribonuclease P
ROP	retinopathy of prematurity
SAP	sphingolipid activator protein
snRNA	small nuclear RNA
snRNP	small nuclear ribonucleoprotein
STRSV	satellite RNA of tobacco ringspot virus
ss	single stranded
TRiC	TCP ring complex
VLBW	very low birth weight
YADH	yeast alcohol dehydrogenase

Bacterial DD-transpeptidases and penicillin

Marc Jamin, Jean-Marc Wilkin and Jean-Marie Frère*

Laboratoire d'Enzymologie and Centre d'Ingénierie des Protéines, Institut de Chimie, B6, Université de Liège, B-4000 Sart-Tilman, Liège 1, Belgium.

Introduction: the bacterial DD-transpeptidases and the peptidoglycan

D-Alanyl-D-alanine peptidases (DD-peptidases) are membrane-bound enzymes involved in the synthesis and remodelling of the peptidoglycan (or murein), a macromolecular sacculus composed of linear glycan chains cross-linked by short peptides (Figure 1), which completely surrounds the cytoplasmic membrane of bacterial cells and is responsible for their shape and mechanical resistance to their own osmotic pressure[1].

The peptidoglycan is a dynamic structure that is continuously remodelled during the cell cycle under the regulated control of two conflicting synthetic (transpeptidase and transglycosylase activities) and hydrolytic (endopeptidase, carboxypeptidase and glycosidase activities) machineries. While the external face of the murein shell is eroded by these autolytic enzymes, disaccharide–peptide precursors linked to an isoprenoid lipid carrier are formed in the cytoplasm, either by *de novo* synthesis or by recycling the liberated peptides. These building blocks are translocated across the cell membrane and incorporated into the peptidoglycan by reactions which occur in the extracellular compartment; disaccharide–peptide units are added to the growing glycan chains and the peptide bridges are subsequently formed by transpeptidation between the peptide chains of adjacent strands. The R-D-alanyl group

To whom correspondence should be addressed.

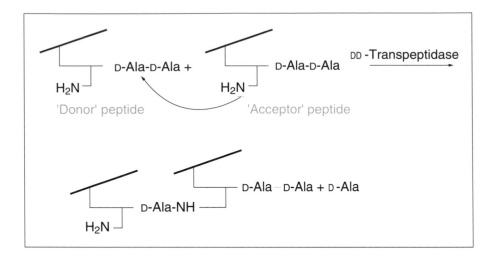

Figure 1. Formation of the peptide bridge in peptidoglycan synthesis
The heavy lines represent the glycan chains composed of alternating N-acetylglucosaminyl and N-acetylmuramyl residues. The peptide moiety attached to the lactyl side-chain of the latter exhibits specific variations in the different bacterial genera. In Gram-negative bacteria, a

$$\text{L-Ala} - \text{D-Glu} \xrightarrow{\quad\quad} \text{m-A}_2\text{pm} - \text{D-Ala–D-Ala}$$
$$\gamma$$

sequence is found, where the free 'acceptor' amino group is on the D centre of meso-diaminopimelic acid (m-A$_2$pm; a detailed structure of the C-terminal tetrapeptide can be found in Figure 3b). For more details, see references 1 and 2 and references therein.

The reaction can also be represented by the simple scheme:

$$R_1\text{-D-Ala-D-Ala} + R_2\text{-NH}_2 \rightarrow R_1\text{-D-Ala-NH-R}_2 + \text{D-Ala}$$

where R$_1$-D-Ala-D-Ala and R$_2$-NH$_2$ are the donor and acceptor peptides, respectively.

of a D-alanyl-D-alanine-terminated 'donor' precursor is transferred to the amino-group of a neighbouring 'acceptor' peptide and the C-terminal D-alanine of the former is released (Figure 1). The equilibrium of that reaction is displaced to the formation of the murein by the insolubilization of the polymer and the diffusion of D-alanine away from the reaction site[2].

How penicillin kills bacteria

All DD-transpeptidases discovered so far are active-site serine enzymes whose catalytic pathways involve transient acylenzyme adducts (Figure 2a) where the penultimate D-alanine residue of the 'donor' substrate is ester-linked to the active-site serine side-chain.

Penicillins, cephalosporins (Figure 2b) and other β-lactam antibiotics inhibit peptidoglycan biosynthesis by inactivating the DD-transpeptidases; they form a covalent, stable acylenzyme with the same residue (Figure 2), thus blocking the bacterial growth and division[3,4]. Uncross-linked peptidoglycan is unable to resist the cell osmotic pressure and lysis most often occurs, but a triggering of the bacterial autolytic system also appears to play an important role in this phenomenon, at least in some species. The specificity of β-lactams as antibacterial agents results from the fact that the peptide moiety of peptidoglycan is unique to the bacterial world, and that no similar transpeptidation reaction involving D-alanyl-D-alanine-terminated peptides exists in eukaryotic organisms.

Figure 2. Catalytic pathway of DD-transpeptidases (E-OH) and inactivation by penicillins (a) and structures of penicillins and cephalosporins (b)

(a) Hydrolysis or aminolysis of the penicilloyl-enzyme is so slow that this reaction is generally devoid of physiological importance. It can also involve an additional breakdown of the penicilloyl moiety. (b) The structures of penicillins and cephalosporins are shown in detail. The β-lactam ring is shown in blue. Typical examples are (c) benzylpenicillin and (d) cephalosporin C. In 6-amino-penicillanate, the amino group of the side-chain is unsubstituted (R_3-CO- is replaced by H).

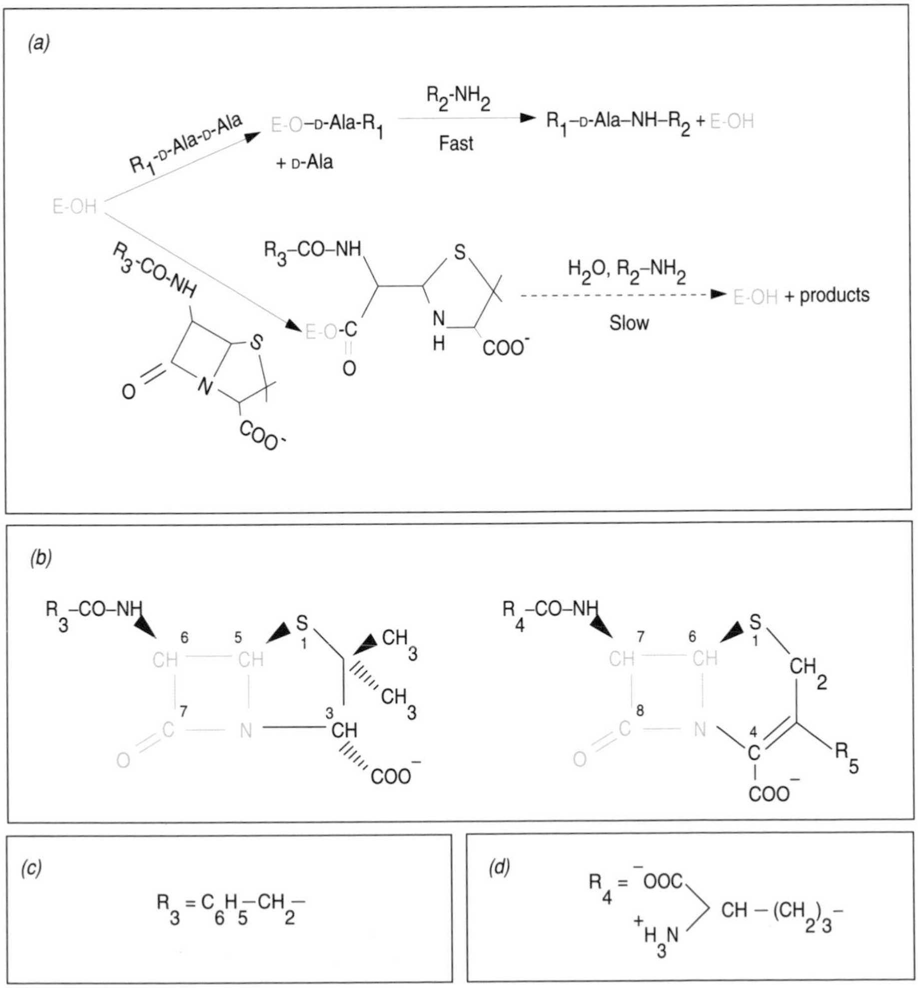

The physiological DD-transpeptidases

The physiologically important DD-transpeptidases are membrane-bound proteins which can be labelled with radioactive or fluorescent penicillins and separated by SDS/PAGE. These techniques reveal the presence of several 'penicillin-binding proteins' (PBPs) in the membrane of all eubacteria[5]. The role of these various enzymes in peptidoglycan synthesis and cell division is presently best understood in *Escherichia coli* (Table 1). The position of the active-site serine in the sequence allowed the PBPs to be divided into low- and high-molecular-mass enzymes. The low-molecular-mass PBPs (LMM-PBPs) exhibit carboxypeptidase (PBPs 4, 5 and 6) activities which seem to be dispensable to the survival of the bacteria but which take part in the regulation of the cell cycle. The high-molecular-mass PBPs (HMM-PBPs) are two-domain proteins with a C-terminal, penicillin-binding domain responsible for the transpeptidase activity. Genetic and morphological approaches highlighted the roles of the HMM-PBPs, and of the products of additional genes, in cell wall elongation, septum formation during the cell division and in shape determination[2,5–7]. Intimately associated with the autolysins in the regulation of the cell cycle, the different DD-transpeptidases undergo activation and deactivation in the different steps of this cycle[8]. However, the catalytic and regulation mechanisms of these enzymes remain poorly understood, since large quantities of purified proteins have never been available.

The cloning and sequencing of the genes coding for various PBPs of several species have supplied detailed information on the primary structures of the corresponding proteins and, more recently, the elimination of the DNA region coding for the membrane-anchoring peptide has allowed the production of some apparently functional, soluble penicillin-binding domains of these enzymes. Nevertheless, most of the information which has been accumulated on the catalytic properties of penicillin-sensitive DD-transpeptidases, and on their interactions with β-lactams, has been obtained with soluble DD-peptidases synthesized by some members of the Actinomycetales order.

The DD-peptidases of *Streptomyces* R61, *Streptomyces* K15 and *Actinomadura* R39[2]

The DD-peptidases of *Streptomyces* R61 and *Actinomadura* R39 are secreted in the extracellular medium as soluble proteins, and that of *Streptomyces* K15 is loosely bound to the cytoplasmic membrane and can be solubilized in the presence of 0.5 M NaCl. The purified proteins catalyse the cleavage of the C-terminal D-alanine of simple synthetic peptides, such as N^α, N^ε-diacetyl-L-lysyl-D-alanyl-D-alanine (Ac$_2$KAA) and N^α-acetyl-L-lysyl-D-alanyl-D-alanine (AcKAA). In the presence of acceptor compounds exhibiting a suitably located amino group, they also perform transpeptidation reactions according to the scheme depicted in the legend of Figure 1. The *Streptomyces* K15 enzyme is so efficient as a transpeptidase that it only hydrolyses a minor proportion of the

Table 1. The PBPs of E. coli

PBP	Approximate molecular mass (kDa)	Name(s) and location of the gene	Physiological roles	Enzymic activities
HMM-PBPs				
1a	92	ponA (mrcA) 75.9 min	Wall elongation	Transpeptidase and transglycosylase
1b	90	ponB (mrcB) 3.5 min	Wall elongation	Transpeptidase and transglycosylase
2	66	pbpA (mrdA) 14.5 min	Shape determination[a]	Transpeptidase
3	60	ftsl (pbpB) 2.0 min	Septation[a]	Transpeptidase
LMM-PBPs				
4	44	dacB 71.7 min	Maturation and recycling	Carboxypeptidase and transpeptidase
5	42	dacA 14.4 min	Regulation of cross-linking	Carboxypeptidase
6	40	dacC 18.0 min	Regulation of cross-linking	Carboxypeptidase

[a]The products of several other genes are involved in the physiological process. Examples of such genes are rodA (14.4 min) and ftsW (2.1 min) for shape determination and septation, respectively.

donor peptide and utilizes the liberated D-alanine as an acceptor in a cyclic, apparently non-productive, reaction which can be detected by adding a small amount of labelled D-alanine to the assay mixture:

$$R\text{-}D\text{-}Ala\text{-}D\text{-}Ala + D\text{-}Ala^* \rightarrow R\text{-}D\text{-}Ala\text{-}D\text{-}Ala^* + D\text{-}Ala$$

The specificity of the enzymes for the acceptors reflects the structure of the strains' peptidoglycan cross-bridges, and, in consequence, the specificity of the physiological transpeptidases. R61 and K15 are *Streptomyces* strains and, thus, the cross-bridge is formed by a

$$\overset{X}{\underset{\underset{(L)}{|}}{}}$$

$$-\text{D-Ala} \longrightarrow \text{Gly} - \text{NH} - \text{CH} - \text{COO}^-$$

sequence and dipeptides, such as Gly-L-Ala, are good acceptors. By contrast, the *Actinomadura* peptidoglycan contains a direct

$$\overset{X}{\underset{\underset{(D)}{|}}{}}$$

$$-\text{D-Ala} \longrightarrow \text{NH} - \text{CH} - \text{COO}^-$$

cross-link (as in Gram-negative bacteria), and only D-amino acids exhibiting a free carboxylate can act as acceptors (see Figure 3b). Both the *Streptomyces* R61 and *Actinomadura* R39 enzymes catalyse the formation of peptide dimers when supplied with adequate acceptor–donor peptides, thus exactly mimicking the transpeptidation reaction (Figure 3). The genes coding for these proteins have been cloned and sequenced. Expression vectors have been constructed which allow their overproduction, and the proteins are now available in large quantities. At present, the *Streptomyces* R61 DD-peptidase is the only penicillin-sensitive enzyme for which a three-dimensional structure has been determined[9] (see below).

Thanks to these advantages, and although the physiological roles of these soluble enzymes are not known, they have been used as model enzymes and several important conclusions have been drawn from the studies of their interactions with penicillins or cephalosporins[10].

- Kinetically, these interactions were characterized by a three-step pathway

$$E + C \underset{}{\overset{K}{\rightleftharpoons}} EC \xrightarrow{k_2} EC^* \xrightarrow{k_3} E + P \text{ (s)}$$

where E is the enzyme, C is the β-lactam, EC is a non-covalent Henri –Michaelis complex (exhibiting a dissociation constant K) and EC* is the acylenzyme.

- The values of k_3 were generally so small (10^{-4} to 10^{-6} s^{-1}) that they were probably physiologically irrelevant. Conversely, the values of K were so large (0.1 to 10 mM) that the sensitivity of the enzyme to a given compound was best characterized by the k_2/K ratio, a second-order rate constant.

- The acylenzyme (EC*) was identified as the penicilloyl- or cephalosporoyl-ester of a serine residue (see Figure 2).

These conclusions were later extended to all other penicillin-sensitive DD-peptidases and PBPs, although the values of k_3 and K could be, respectively, as high as 5×10^{-3} s^{-1} or as low as 5 µM in a few exceptional cases. These properties allowed penicillins to be utilized as active-site titrating agents, and this type of experiment represents the best purity criterion for these enzymes.

DD-Peptidases and β-lactamases

Parallel studies performed with β-lactamases[11] — extracellular enzymes which efficiently hydrolyse the β-lactam ring and represent the most widespread factors in the resistance of bacteria to β-lactams — demonstrated that most of these were also active-site serine enzymes, utilizing the same three-step catalytic pathway as DD-peptidases and forming similar transient acylenzymes (several Zn^{2+}-dependent β-lactamases have also been identified but they will not be discussed here). However, the k_2 and, more strikingly, the k_3 values were much larger with the β-lactamases (up to several thousand per second). On the basis of their primary structures, serine β-lactamases have been divided into three classes: A, C and D (class B contains the Zn^{2+}-β-lactamases). Sequence comparisons with the DD-peptidases, performed with the help of the

Figure 3. Peptides used as donor–acceptor substrates by the *Streptomyces* R61 (*a*) and *Actinomadura* R39 (*b*) DD-peptidases

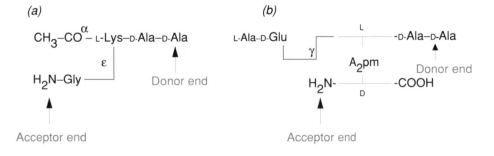

usual algorithms, failed to reveal statistically significant analogies which could indicate a divergent evolution from a common ancestor. Surprisingly, the structural data obtained by X-ray crystallography[12–15], indicated an evident similarity between the *Streptomyces* R61 DD-peptidase and the β-lactamases (Figure 4) and demonstrated the presence, in the immediate vicinity of the active-site residues, of three conserved structural elements occupying equivalent positions and exhibiting identical or similar functional groups[16].

- Element 1: a Ser-Xaa-Xaa-Lys sequence, where Ser is the active serine, situated at the *N*-terminus of a rather long and hydrophobic helix, so that the side-chain of the lysine residue, just one turn after the Ser, also lies in the active site.

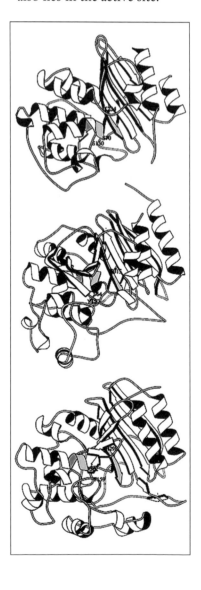

- Element 2: A Tyr-Ser-Asn triad in the *Streptomyces* R61 enzyme, on a loop forming one side of the catalytic cavity, with the tyrosine and asparagine side-chains pointing into this cavity and corresponding to Tyr-Xaa-Asn, in the class C and D β-lactamases, and to Ser-Xaa-Asn, in class A β-lactamases and most PBPs.

- Element 3: A His-Thr-Gly triad, on a piece of β-strand, forming the opposite wall of the active site and corresponding to Lys-Thr(Ser)-Gly sequences in the β-lactamases and Lys-Thr-Gly sequences in all other PBPs.

Figure 4. Structures of class A (top) and class C (middle) β-lactamases and of the *Streptomyces* R61 DD-peptidase (bottom)

It must, however, be emphasized that, at the present stage of knowledge, it cannot be concluded that these chemically similar functional groups necessarily play the same roles in the catalytic mechanisms of the various enzymes. Indeed, if the identity of the active serine is clearly established in all cases, the only residue whose universal conservation is well-understood is the glycine residue of the third element — since the presence of any side-chain in this position would sterically hinder the interaction between the active-site serine and the substrates.

Although no 3-dimensional structural data are available for other PBPs, these structural elements (with minor variations) could be located in all the available sequences, and it can be safely predicted that their spatial dispositions will be closely related to those in the model *Streptomyces* R61 enzyme.

Ester and thiolester substrates

The simple tri-, tetra- and pentapeptide substrates of the *Streptomyces* R61, *Streptomyces* K15 and *Actinomadura* R39 DD-peptidases, described above in the text and in Figure 3, are recognized neither by the β-lactamases nor by the physiologically important HMM-PBPs. Some of the latter catalyse coupled transglycosylase and transpeptidase reactions when supplied with the natural, lipid-linked, disaccharide–peptide precursor, but this experimental system is so complex that only qualitative results can be obtained. Careful studies were performed of the concomitant hydrolysis and transpeptidation reactions catalysed by the *Streptomyces* R61 enzyme in the presence of Ac$_2$KAA, as donor substrate, and of various acceptors; however, the assays, based on discontinuous measurements of the released D-alanine or of the radioactive products and residual tripeptide after electrophoretic separation, were rather painstaking[17]. Moreover, acylation appeared to be rate-limiting, so that little information could be gathered about the subsequent steps. Nonetheless, the variations of the transpeptidation/hydrolysis ratio with the donor and acceptor concentrations suggested the prevalence of a complex mechanism, involving more than one binding site for each substrate. In this context, the introduction of new ester and thiolester substrates (Table 2) represented a major advance, with both conceptual and practical implications[18,19]. First, some of these compounds could be utilized by the DD-peptidases, β-lactamases and even by some HMM-PBPs in hydrolysis and/or transacylation reactions, underlining an additional, functional relationship between these different proteins. (Interestingly, class C β-lactamases were found to catalyse transacylation reactions much more efficiently than their class A and D counterparts.) Secondly, the disappearance of the ester bond in substrate III and of the thiolester bond in substrates V–VII (see Table 2) could be directly visualized by spectrophotometry, allowing a continuous monitoring of the DD-trans- and carboxypeptidase activities for the first time. Moreover, with the ester and thiolester substrates, deacylation was found to be rate-limiting with several enzymes, so that the complete catalytic

Table 2. Structures and properties of the peptide, ester and thiolester substrates

	Structure	Substrate for
I	CH_3 CH_3 Ac_2-L-Lys-NH-CH-CO-NH-CH-COO⁻ (D) (D)	DD-peptidases
II	CH_3 CH_3 Ac_2-L-Lys-NH-CH-CO-O-CH-COO⁻ (D) (D)	DD-peptidases
III	CH_2-C_6H_5 C_6H_5-CO-NH-CH_2-CO-O-CH-COO⁻ (D)	DD-peptidases β-lactamases
IV	CH_3 CH_3 C_6H_5-CH_2-CO-NH-CH-CO-O-CH-COO⁻ (D) (D)	DD-peptidases β-lactamases
V	C_6H_5-CO-NH-CH_2-CO-S-CH_2-COO⁻	DD-peptidases β-lactamases
VI	CH_3 C_6H_5-CO-NH-CH-CO-S-CH_2-COO⁻ (D)	DD-peptidases β-lactamases HMM PBPs
VII	CH_3 CH_3 C_6H_5-CO-NH-CH-CO-S-CH-COO⁻ (D) (D)	DD-peptidases β-lactamases HMM PBPs

pathway could be more easily studied[20]. A detailed analysis of the interactions between the *Streptomyces* R61 enzyme, substrate V and various acceptors led to the minimum kinetic model depicted by Figure 5, which rested on the following observations[21].

- In the absence of acceptor, the hydrolysis reaction obeyed the Henri–Michaelis equation, even at high substrate concentrations ($[S_1] > K_m$).

- The transpeptidation/hydrolysis ratio was not strictly proportional to the acceptor concentration, but reached a limiting value, which indicated that hydrolysis could occur from an acceptor-containing intermediate.

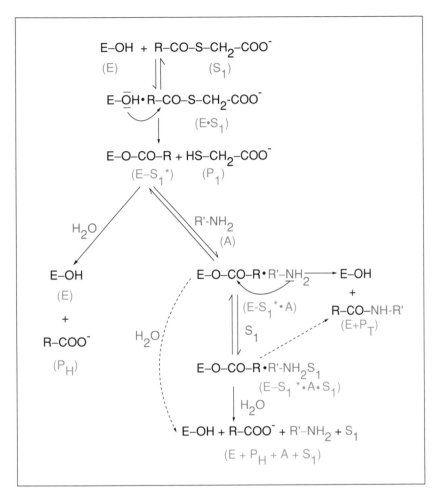

Figure 5. Kinetic model for the concomitant transfer and hydrolysis reactions catalysed by the _Streptomyces_ R61 DD-peptidase
The reactions shown as dashed lines might occur but are not necessary in a 'minimum' scheme. S_1 and A are, respectively, the donor and acceptor substrates. P_1, P_T and P_H are the products of the reaction: the leaving group, the transacylation and the hydrolysis products, respectively.

- The same ratio varied with the donor concentration, which implied either that the transpeptidation pathway first involved the binding of the acceptor to the free enzyme or that the donor could bind to the acceptor-containing intermediate. The first possibility could be rejected because no binding of simple acceptors, such as D-alanine, to the free enzyme could be detected.

- The presence of the acceptor did not modify the $k_{cat.}/K_m$ ratio measured in the hydrolysis reaction, indicating that the formation of a catalytically competent enzyme–donor–acceptor intermediate only occurred after the formation of the acylenzyme ES*.

The same model not only nicely accounted for the results previously obtained with the tripeptide substrate as a donor, but also appeared to prevail for a soluble form of PBP2X, one of the physiologically essential PBPs of *Streptococcus pneumoniae*, which was obtained as a soluble protein by deleting the part of the gene coding for its membrane-anchoring peptide. It failed to recognize the Ac$_2$KAA and AcKAA tripeptides, but efficiently hydrolysed the thiolester VI and, at present, remains the only HMM-PBP for which a kinetic study of the transpeptidation reaction has been performed[22]. The presence of two distinct binding sites for the donor substrate might seem surprising, but the situation is even more complex. Indeed, with the *Streptomyces* R61 enzyme, acceptors such as the dipeptides Gly-L-Ala and Gly-L-Gln, whose structures more closely mimic that of the natural substrate, were found to inhibit the total (transpeptidation + hydrolysis) reaction at high concentrations.

With the *Actinomadura* R39 enzyme, an excess of the 'natural' acceptor tetrapeptide

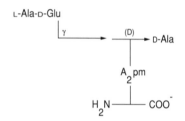

induced a complete 'freezing' of the system[23]. This suggested the existence of a second, inhibitory acceptor site. It could be hypothesized that this second acceptor site partly or completely overlaps with the second donor site proposed in Figure 5, but the occupation of this site by an acceptor molecule would result in the formation of a 'dead-end' complex. An additional argument in favour of the existence of this second acceptor site is the fact that, with the *Streptomyces* R61 enzyme, the rate of acylation by penicillins is modified by the 'inhibitory' acceptors, but not by the simple ones, such as D-alanine.

The 'minimum' model of Figure 5 might not be completely valid for all enzymes. Indeed, with the *Actinomadura* R39 and *Streptomyces* K15 enzymes, and with the class C β-lactamases, the acceptors modified the $k_{cat.}/K_m$ value observed for the hydrolysis reaction, a result which can only be interpreted by assuming the binding of the acceptor before the formation of the acylenzyme ES*, i.e. to the Henri–Michaelis intermediate ES. One might wonder about the relevance of these complex models for the physiological transpeptidation reaction: could the utilization of small substrates result in artefacts which would disappear with the larger natural peptides? Although this hypothesis cannot be completely neglected, it is striking to remember that the inhibitory acceptors are those most similar to the natural ones. The existence of secondary, maybe regulatory, sites on the DD-transpeptidases thus appears as a distinct physiologically important possibility.

Modification of the *Streptomyces* R61 enzyme by site-directed mutagenesis[24-28]

The 10 residues composing the three structural elements identified in the primary and tertiary structures are shown in Figure 6, where a molecule of cephalosporin C has been positioned by molecular modelling. Among these 10 residues, Val-63, Thr-64, Ser-160 and Gly-300 were not modified. On the basis of the preliminary X-ray data, the first three were supposed to point away from the catalytic cavity and to be mainly involved in maintaining the general architecture. Similarly, any side-chain on the third residue of element 3 (Gly-300) would protrude into the active site and severely hinder the approach of the substrate. The studies summarized in Table 3 have confirmed the crucial role played by the six other residues, although their exact functions are not necessarily understood.

The peptidase activity does not tolerate modifications

The most striking results were those obtained with the peptide substrate Ac_2KAA and indicated that the acylation step, which is already rate-limiting ($k_2 < k_3$) with the wild-type enzyme, was extremely sensitive to any modification of the active-site side-chains. This underlined the need for a very accurate orientation of the catalytic machinery for optimal interaction with the close analogues of the natural substrates. The efficiency of the deacylation step, which could only be probed with the thiolesters, appeared to be much more resilient, and even remained very good in several cases. Although it might be dangerous to extrapolate these data to the peptide, which has a very different and much more specific acyl moiety, it was clear that the deacylation step never became rate-limiting with this latter substrate.

The thiolesterase activity and the second element

The rates of acylation of some modified enzymes by the thiolester remained high. This could be explained, in part, by the fact that the $^-S–CH_2–COO^-$ group is a much better leaving group than D-alanine, as illustrated by the strongly increased intrinsic reactivity of the thiolester. It would be tempting to conclude that the side-chains of Tyr-159 and, to a lesser degree, His-298 and Thr-299 are involved in the protonation of this leaving group, an hypothesis which is, however, somewhat undermined by the behaviour of the Tyr-159 → Phe enzyme. It is also interesting to note that the acylation of the Tyr-159 → His and Tyr-159 → Ser mutants by the thiolester was significantly less impaired than the deacylation. The properties of the Tyr-159 → Phe mutant were much more uniform, underlining the capacity to form hydrogen bonds as a major factor in the large residual thiolesterase activity of Tyr-159 → His and Tyr-159 → Ser. The same two mutants also exhibited sensitivities to penicillins and cephalosporins similar to, or even higher than, those of the wild-type protein. Although these β-lactams are certainly more reactive

Table 3. Acylation and deacylation parameters and transpeptidation/hydrolysis (T/H) ratios for the various mutant proteins (given as percentage of wild-type values)

	k_2/K: acylation				k_3: deacylation	
	Peptide	Thiolester V	Benzylpenicillin	Cephalosporin C	T/H [a]	Thiolester V
Element 1 (Ser-62-Val-Thr-Lys)						
Ser-62 → Cys	0	0	0	0	-	-
Lys-65 → Arg	0.5	0.6	<0.01	ND	ND	>100
Element 2 (Tyr-159-Ser-Asn)						
Tyr-159 → His	0.4	48	500	90	4	3
Tyr-159 → Ser	0.3	60	55	35	4	6
Tyr-159 → Phe	0.1	0.1	0.3	1	6	0.2
Asn-161 → Ser	3	38	80	33	15	7
Asn-161 → Ala	0.3	3.2	1.3	2.2	30	44
Element 3 (His-298-Thr-Gly)						
His-298 → Lys	1	44	1.3	8	7.5	20
His-298 → Gln	0.6	10	0.65	2	22	30
Thr-299 → Val	0.15	9	3.3	<0.01	1.5	36
Tyr-159 → Ser/His-298 → Lys	0.01	1.1	1.6	0.5	0	1.6

The values lower than 1% are shown in bold type. [a]Transpeptidation/hydrolysis ratio with thiolester V and D-alanine.

than the peptide, they are also orders of magnitude more stable than the thio-lester, and it is clear that the observed rapid acylation requires a compensating mechanism for the protonation of the leaving group. In this respect, an imida-zole side-chain might supply a suitable replacement for the phenol group, and one or more water molecules could occupy the 'hole' created by the Tyr-159 → Ser mutation. Again, both these alternative mechanisms would be impossible for the Tyr-159 → Phe mutant. The side-chain Asn-161 might form a hydrogen bond with the carbonyl oxygen of the 6(7)-β-acylamido side-chain of penicillins (or cephalosporins), the lysyl residue of the peptide or the ben-zoyl group of the thiolester substrates. Its replacement by a serine residue only significantly affected acylation by the peptide and, to a lesser degree, the hydrolysis of the benzoylglycyl adduct formed with the thiolester. Surprisingly, this latter reaction was less impaired by the Asn-161 → Ala sub-

Figure 6. Detailed structure of the three conserved elements (SVTK, YSN and HTG) forming the active site of the *Streptomyces* R61 DD-peptidase, in which a molecule of cephalosporin C has been 'docked' by computer modelling

The antibiotic molecule is shown in light blue, the protein backbone in grey and the α-carbon atoms are labelled blue circles. The only represented hydrogen atoms are those of the backbone amide groups.

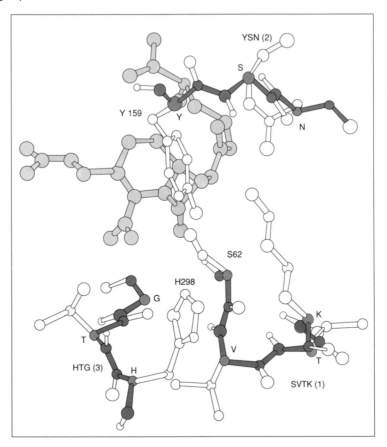

stitution which, conversely, severely decreased all acylation rates. These observations underlined the necessity for a precise positioning of the acylating agents in the enzyme catalytic site and the role of the hydrogen-bonding properties of the side-chain of residue 161 in this positioning. Understandably, this role would be much less crucial in the deacylation step.

The first element cannot be modified

Although a strong impairment was expected for the Ser-62 → Cys mutant, its magnitude was quite surprising. Indeed, similarly modified class A and class C β-lactamases retained a non-negligible proportion of activity and, similarly, thiolsubtilisin is reasonably active, at least on its less-specific substrates. Here, no acylation was detected, even with the thiolester V, which could be considered as the equivalent of *p*-nitrophenyl acetate for subtilisin. The serine hydroxyl side-chain thus appears to be a consistent prerequisite for the DD-peptidase and penicillin-binding activities. The lysyl side-chain of the same element is conserved in all penicillin-recognizing enzymes, both PBPs and β-lactamases. The relatively conservative lysine to arginine mutation resulted in spectacular decreases of all the activities, but it was the only case where the acylation rate was affected more significantly with penicillin than with the substrates. The modified enzyme thus behaved as a poor, but penicillin-insensitive, DD-peptidase where the deacylation rate was equal to or larger than that recorded with the wild-type protein.

In β-lactamases, the major consequences of the corresponding Lys-73 → Arg (class A) and Lys-67 → Arg (class C) mutations resulted in marked decreases of the acylation rates, but this was apparently due to a decrease of k_2 and an increase of K in class A and class C, respectively. The k_3 values were less, or not, modified and even increased for some poor substrates of the class C enzyme[29,30].

The third element

The threonine residue of the third element is also totally conserved in all PBPs and the hydroxyl function is present in all β-lactamases since the only known substitution is threonine to serine in some of these latter enzymes. The disappearance of this function severely impaired the DD-peptidase, but here one of the most unexpected observations was the complete disappearance of the cephalosporin-binding properties, which contrasted with a relatively rapid acylation by penicillins. This behaviour, unique among the DD-peptidase mutants, is reminiscent of the effects of the Asn-132 → Ser mutation in a class A β-lactamase[32]. Paradoxically, this latter substitution is located in element 2 and the equivalent mutation in the DD-peptidase has no such effects. This observation highlights the distinct behaviours of the two types of enzyme despite the structural similarities outlined above.

The *Streptomyces* R61 enzyme is the only known PBP where the first residue of element 3 is a histidine residue[16]. The His-298 → Lys mutation, which generates a Lys-Thr-Gly triad identical to that found in other PBPs, strongly decreased the activity. Replacements of the corresponding lysine residue of several β-lactamases and of *E. coli* PBP5 have shown that a positive charge on this side-chain is a necessary, but not always sufficient, condition for an optimum activity. It would thus be tempting to conclude that the p*K* of 9.0–9.5, observed in the dependencies of the rates of acylation by the peptide and the antibiotics, is that of the imidazole side-chain, significantly increased by the active-site environment. Indeed, the lysine to histidine mutation in the *Streptomyces albus* G class A β-lactamase resulted in the appearance of a new p*K* of 6.5, which could be attributed to the newly introduced histidine residue[31]. Each enzyme has thus adopted a structure such that optimum activity is obtained with different residues in this position.

Transpeptidation strongly relies on elements 2 and 3

All the modifications in elements 2 and 3 specifically decreased the transpeptidation/hydrolysis ratio. Tyr-159 and Thr-299 appeared to be most strongly involved in this phenomenon, but the complete loss of transpeptidation properties for the double Tyr-159 → Ser + His-298 → Lys mutant underlined a cooperation between the two elements in the catalysis of this reaction. The three modified residues must be involved in the proper positioning, or in the activation, of the acceptor substrate, but other residues situated in the same part of the protein might also contribute to the formation of an efficient acceptor site.

The search for a general base

In class A β-lactamases, residue Glu-166 — situated between elements 2 (Ser-130-Asp-Asn) and 3 (Lys-234-Thr-Gly) — is thought to act as a general base involved in the activation of the Ser-70 residue in acylation and of a water molecule in deacylation[29,31–33]. Another hypothesis would attribute this role to the lysine residue of element 1, which would exhibit an unusually low p*K* value of 4.5–5.0[34]. In the structure of class C enzymes, no carboxylate group is found in an equivalent position and it has been proposed that Tyr-150 of element 2, in the deprotonated phenolate form, would play the same role[14]. In the *Streptomyces* R61 enzyme, all the aspartate and glutamate residues between positions 162 and 297 whose side-chain might, on the basis of the α-carbon trace, lie in the vicinity of the active site have been replaced by non-acidic residues. None of these mutations has resulted in significant impairments of the enzyme activity and the authors are forced to conclude that the general base must originate from another segment of the sequence (J.-M. Wilkin, B. Joris and J.-M. Frère, unpublished work). His-298 does not seem to be suitably situated to fulfil this function and the fact that it is replaced by a lysine residue in other PBPs, as well as the behaviour of the lysine to histidine mutant of the β-lactamase described above, argue against this possibility. Lys-65 and Tyr-159 remain possible candidates and in the next section mechanisms are proposed and criticized on the basis of the available data.

Hypothetical mechanisms

Figure 7 details two possible mechanisms for the acylation reaction. In both of them, and by analogy with the models widely accepted for the β-lacta-mases[29,33], the carbonyl oxygen of the substrate scissile peptide bond (or of the β-lactam ring) forms hydrogen bonds with the peptide backbone NH-groups of Ser-62 and Ser-301, which behave as an oxyanion hole. In the first mechanism, an unidentified general base (Lys-65?) accepts the proton from the active serine during the formation of the tetrahedral intermediate, whose breakdown is catalysed by the protonation of the leaving group by Tyr-159. During this

Figure 7. Mechanisms of acylation involving (a) an unidentified group B or (b) Tyr-159 as general bases

The substrate carbonyl group is polarized by interaction with the backbone amide groups of Ser-62 and Thr-301, forming the oxyanion hole. Tyr-159 donates a proton to the leaving group. (a) No phenolate ion is involved, since a proton is back-delivered to Tyr-159, originating from the acidic form of the general base B and via additional residues and/or water molecules, as proposed by Lamotte-Brasseur et al.[33] (b) The phenolate ion, already present in the free enzyme, would transfer the Ser-62 proton to the leaving group, acting as the equivalent of His-57 in chymotrypsin. In both cases, deacylation would occur by a reversal of the acylation steps, after replacement of the leaving group by a water molecule (hydrolysis) or a new aminated acceptor (transpeptidation).

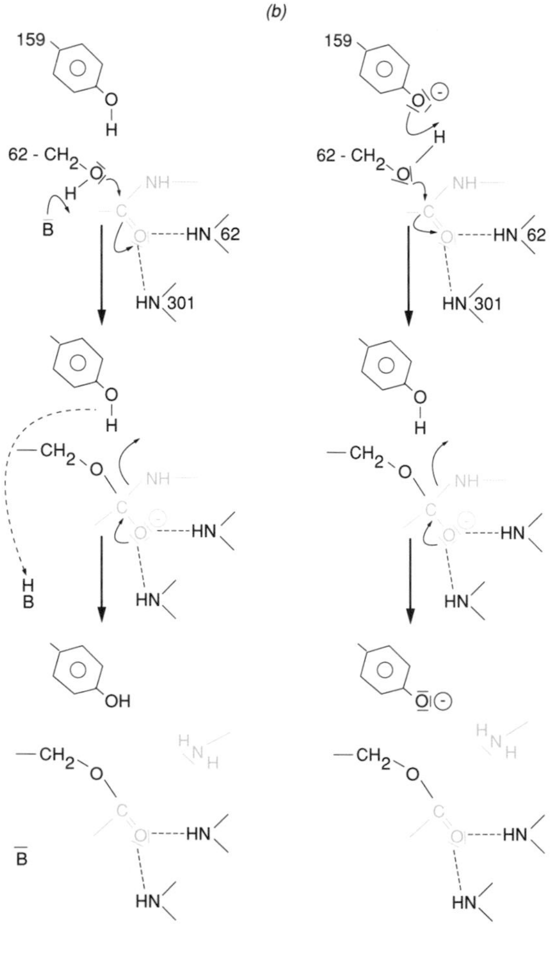

step, the proton must be transferred from the protonated base (BH) to the phenolate ion of Tyr-159. This could be performed by a proton-relay system similar to that proposed for the class A β-lactamase of *S. albus G*[33], but the residues involved in this transfer remain to be identified. In the second mechanism, Tyr-159 itself acts as the general base. In both cases, deacylation would occur according to the reverse pathway, with water or the acceptor replacing the leaving group.

Both these hypotheses suffer several drawbacks. The first rests on a mysterious general base, and the transfer of the proton from BH to the phenolate remains just as hypothetical. If Lys-65 is the general base, its pK must be quite low (<5 on the basis of the pH-activity curves) and the efficient deacylation observed with the Lys-65 → Arg mutant is difficult to explain. The second fails to account for the good acylation of the Tyr-159 → Ser mutant by penicillins and cephalosporins. Indeed, if the histidine residue of the Tyr-159 → His mutant might be expected to act as a general base, it is certainly not so for the serine residue of Tyr-159 → Ser. Moreover, the tyrosine pK should also be very significantly decreased. A hybrid mechanism (Figure 8) might solve some

Figure 8. Hybrid mechanism
In the free enzyme, Tyr-159 would remain protonated and the phenolate ion formed at the level of the acylenzyme adduct (3) would then act as the general base in the deacylation step, where the acidic form of the unindentified general base B would back-deliver a proton to Ser-62.

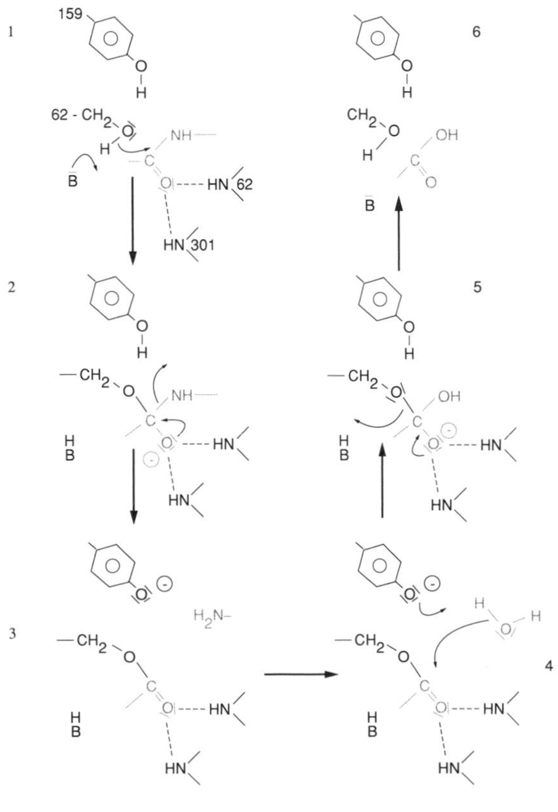

of these difficulties. It would not involve the transfer of the proton from BH to the phenolate during the breakdown of the first tetrahedral intermediate. The phenolate ion would then act as the general base to activate the acceptor or the water molecule in the deacylation, and the protonated general base would deliver the proton back to Ser-62 during the breakdown of the second tetrahedral intermediate. This hypothesis is supported by the fact that the Tyr-159 → His and Tyr-159 → Ser mutations appear to affect the deacylation step more significantly than the acylation when the thiolester V is the substrate. Indeed, the good mercaptoacetate leaving group does not need to be protonated for an efficient acylation to occur. By contrast, the absence of the phenolate ion would then present a more serious obstacle to the deacylation reaction.

Although this might look like an attractive hypothesis, it fails to solve the mystery of the general base. Moreover, in several PBPs, the first residue of the second element is a serine residue, which cannot possibly fulfil the function of the tyrosine phenol side-chain assumed by the hybrid mechanism. We are quite conscious that these constitute major handicaps, unless PBPs have evolved two distinct catalytic mechanisms.

Penicillin-resistant PBPs

The numerous PBPs which have been studied so far exhibit large variations in their sensitivities to β-lactams. For example, with benzylpenicillin, the k_2/K values range from 200 to 300 $M^{-1}\cdot s^{-1}$ for some *E. coli* PBPs to 300000 $M^{-1}\cdot s^{-1}$ for the *Actinomadura* R39 DD-peptidase [2]. Moreover, with the same enzyme, the k_2/K value is also strongly dependent upon the exact structures of the β-lactam molecule side-chain (R_3, R_4 and R_5, Figure 2b). A lower k_2/K value results in a slower acylation rate and, if this PBP is physiologically essential, consequently confers an increased 'intrinsic' resistance to the bacterium — as opposed to that mediated by the production of one or several β-lactamases or, in Gram-negative strains, by a reduction of the outer membrane permeability. It could thus be expected that resistant strains might emerge by producing PBPs exhibiting strongly reduced affinities for penicillins. In theory, this could be realized by acquiring a 'resistant' gene from a different, more resistant, species or by modifying an existing gene. Bacteria have successfully utilized both strategies, but the second appears to present strong limitations. Indeed, successive mutations have generally resulted in marginal decreases of the penicillin sensitivity (factors of at most 10), although some exceptions are known. The most spectacular resistance phenomena rest either on the first strategy, such as the acquisition of PBP2′ by the methicillin-resistant *Staphylococcus aureus*, or on a profound remodelling of the structural genes by the replacement of large oligonucleotide stretches by the corresponding portions of genes coding for resistant proteins. With these proteins, k_2/K values as low as 10–20 $M^{-1}\cdot s^{-1}$ for benzylpenicillin have been determined [2,35]. More resistant

PBPs might also originate from increased k_3 values. Interestingly, this mechanism has not presently been found to occur in resistant strains.

Strikingly, in all low-affinity PBPs, the three conserved elements described above (see page 8) are invariably found in positions similar to those observed in the sequences of their penicillin-sensitive counterparts. The structural origin of the decreased acylation rates by β-lactams thus remains mysterious. Unfortunately, it has not been possible to perform estimations of the DD-transpeptidase activity of these proteins *in vitro*. In some cases, the thiolester VI was utilized[36] but this reaction does not always faithfully reflect the behaviour of the enzyme towards the natural peptide. It can, however, be assumed that these low-affinity PBPs have conserved a sufficient transpeptidase activity to allow them to take over peptidoglycan biosynthesis when more sensitive enzymes are inactivated. The fact that, in some cases, appearance of resistance requires their overproduction, underlines the possibility that they might indeed also be enzymically less efficient[2].

Conclusion: does penicillin behave as a substrate analogue?

In the last 30 years, the study of the mechanism of action of penicillins has been dominated by the very fruitful hypothesis of Tipper and Strominger[37], who argued that penicillins behaved as structural analogues of the natural peptide substrates of the DD-transpeptidases. Although it is now widely recognized that both types of compound do indeed acylate the same active serine residue, which implies similar positioning of the scissile peptide and β-lactam amide bonds in the catalytic cavities, several results indicate this analogy to be far from perfect. The structures of the R_1 and R_3 or R_4 groups (Figure 2*b*) in the substrate and antibiotic, respectively, strongly influence the rate of enzyme acylation; however, the same substitutions can have very different effects on the relative efficiencies of the substrates and of the inactivators. Conversely some of the modified proteins discussed here (Tyr-159 → His and Tyr-159 → Ser) react very efficiently with benzylpenicillin and cephalosporin C but barely recognize the 'good' peptide substrates of the wild-type enzyme. Their behaviours make them somewhat β-lactamase-like, (at least as far as acylation is concerned), since these latter enzymes fail to interact with D-Ala-D-Ala-terminated peptides. Similarly, although 6-aminopenicillanate, which is devoid of side-chain, remains a good substrate for several β-lactamases, it reacts very slowly with most PBPs and does not kill bacteria, at least at clinically useful concentrations. Finally, the penicillin-resistant PBPs represent the other extreme, i.e. proteins which can nearly escape the acylating action of β-lactams, but still possess a DD-transpeptidase activity, sufficient to allow the cells to survive. Only a detailed knowledge of the catalytic cavity structures of various PBPs, exhibiting different sensitivities to penicillins, combined with computer-aided modelling of the enzyme–substrate and enzyme–inactivator interactions will supply a final evaluation of the exact degree of structural analogy that a potential inactivator must exhibit so that the enzyme can be fooled into welcoming this traitor into its precious active-site cavity.

The authors wish to thank Dr J.R. Knox and Dr J.A. Kelly (University of Connecticut, U.S.A.) for the drawings of the enzyme structures. The work in Liège was supported by the Belgian Programme on Inter-University Poles of Attractions, by Actions Concertées with the Belgian government and various contracts with the Fonds National de la Recherche Scientifique (Brussels). JMW and MJ both benefited from IRSIA fellowships between 1988 and 1993.

References

1. Frère, J.M. & Joris, B. (1985) Penicillin-sensitive enzymes in peptidoglycan biosynthesis. *CRC Crit. Rev. Microbiol.* **11**, 299–396
2. Frère, J.M., Nguyen-Distèche, M., Coyette, J. & Joris, B. (1992) Mode of action: interaction with the penicillin-binding proteins, in *The Chemistry of Beta-lactams* (Page, M.I., ed.), pp. 148–195, Chapman and Hall, Glasgow
3. Frère, J.M., Duez, C., Ghuysen, J.M. & Vandekerckhove, J. (1976) Occurrence of a serine residue in the penicillin-binding site of the exocellular DD-carboxy-peptidase-transpeptidase from *Streptomyces* R61. *FEBS Lett.* **70**, 257–260
4. Yocum, R.R., Amanuma, H., O'Brien, T.A., Waxman, D.J. & Strominger, J.L. (1982) Penicillin is an active site inhibitor for four genera of bacteria. *J. Bacteriol.* **149**, 1150–1153
5. Spratt, B.G. & Pardee, A.B. (1975) Penicillin-binding proteins and cell shape in *E. coli. Nature (London)* **254**, 512–513
6. Spratt, B.G. (1977) Properties of the penicillin-binding proteins of *Escherichia coli* K12. *Eur. J. Biochem.* **72**, 341–352
7. Ghuysen, J.M. (1991) Serine β-lactamases and penicillin-binding proteins. *Annu. Rev. Microbiol.* **45**, 37–67
8. Nanninga, N. (1991) Cell division and peptidoglycan assembly in *Escherichia coli. Mol. Microbiol.* **5**, 791–795
9. Kelly, J.A., Knox, J.R., Zhao, H., Frère, J.M. & Ghuysen, J.M. (1989) Crystallographic mapping of β-lactams bound to a D-Alanyl-D-Alanine peptidase target enzyme. *J. Mol. Biol.* **209**, 281–295
10. Frère, J.M., Ghuysen, J.M. & Iwatsubo, M. (1975) Kinetics of interaction between the exocellular DD-carboxypeptidase-transpeptidase from *Streptomyces* R61 and β-lactam antibiotics. A choice of models. *Eur. J. Biochem.* **57**, 343–351
11. Waley, S.G. (1992) β-lactamases: mechanism of action, in *The Chemistry of Beta-lactams* (Page, M.I., ed.), pp. 198–228, Chapman and Hall, Glasgow
12. Kelly, J.A., Dideberg, O., Charlier, P., Wery, J., Libert, M., Moews, P., Knox, J., Duez, C., Fraipont, C., Joris, B., *et al.* (1986) On the origin of bacterial resistance to penicillin: comparison of a β-lactamase and a penicillin target. *Science* **231**, 1429–1431
13. Samraoui, B., Sutton, B., Todd, R., Artymiuk, P., Waley, S.G. & Phillips, D. (1986) Tertiary structure similarity between a class A β-lactamase and a penicillin-sensitive D-alanyl-carboxypeptidase-transpeptidase. *Nature (London)* **320**, 378–380
14. Oefner, C., Darcy, A., Daly, J.J., Gubernator, K., Charnas, R.L., Heinze, I., Hubschwerlen, C. & Winkler, F.K. (1990) Refined crystal structure of beta-lactamase from *Citrobacter freundii* indicates a mechanism for beta-lactam hydrolysis. *Nature (London)* **343**, 284–288
15. Lobkovsky, E., Moews, P.C., Hansong, L., Zhao, H., Frère, J.M. & Knox, J.R. (1993) Evolution of an enzyme activity: crystallographic structure at 2.8 A resolution of cephalosporinase from ampC

gene of *Enterobacter cloacae* P99 and comparison with a class A penicillinase. *Proc. Natl. Acad. Sci. U.S.A.* **90**, 11257–11261

16. Joris, B., Ledent, P., Dideberg, O., Fonzé, E., Lamotte-Brasseur, J., Kelly, J.A., Ghuysen, J.M. & Frère, J.M. (1991) Comparison of the sequences of Class-A beta-lactamases and of the secondary structure elements of penicillin-recognizing proteins. *Antimicrob. Agents Chemother.* **35**, 2294–2301

17. Frère, J.M., Ghuysen, J.M., Perkins, H.R. & Nieto, M. (1973) Kinetics of concomitant transfer and hydrolysis reactions catalysed by the exocellular DD-carboxypeptidase-transpeptidase of *Streptomyces* R61. *Biochem. J.* **135**, 483–492

18. Pratt, R.F. & Govardhan, C.P. (1984) β-Lactamase-catalysed hydrolysis of acyclic depsipeptides and acyl-transfer to specific amino acid acceptor. *Proc. Natl. Acad. Sci. U.S.A.* **81**, 1302–1306

19. Adam, M., Damblon, C., Plaitin, B., Christiaens, L. & Frère, J.M. (1990) Chromogenic depsipeptide substrates for β-lactamases and penicillin-sensitive DD-peptidases. *Biochem. J.* **270**, 525–529

20. Jamin, M., Adam, M., Damblon, C., Christiaens, L. & Frère, J.M. (1991) Accumulation of acyl-enzyme in DD-peptidase-catalysed reactions with analogues of peptide substrates. *Biochem. J.* **280**, 499–506

21. Jamin, M., Wilkin, J.M. & Frère, J.M. (1993) A new kinetic mechanism for the concomitant hydrolysis and transfer reactions catalysed by bacterial DD-peptidases. *Biochemistry* **32**, 7278–7285

22. Jamin, M., Damblon, C., Millier, S., Hakenbeck, R. & Frère, J.M. (1993) Penicillin-binding protein 2X of *Streptococcus pneumoniae*: enzymic activities and interactions with β-lactams. *Biochem. J.* **292**, 735–741

23. Ghuysen, J.M., Leyh-Bouille, M., Campbell, J.N., Moreno, R., Frère, J.M., Duez, C., Nieto, M. & Perkins, H.R. (1973) Structure of the wall peptidoglycan of *Streptomyces* R39 and the specificity profile of its exocellular DD-carboxypeptidase-transpeptidase for peptide acceptors. *Biochemistry* **12**, 1243–1250

24. Hadonou, A.M., Jamin, M., Adam, M., Joris, B., Dusart, J., Ghuysen, J.M. & Frère, J.M. (1992) Importance of the His-298 residue in the catalytic mechanism of the *Streptomyces* R61 extracellular DD-peptidase. *Biochem. J.* **282**, 495–500

25. Hadonou, A.M., Wilkin, J.M., Varetto, L., Joris, B., Lamotte-Brasseur, J., Klein, D., Duez, C., Ghuysen, J.M. & Frère, J.M. (1992) Site-directed mutagenesis of the *Streptomyces* R61 DD-peptidase: catalytic function of the conserved residues around the active site and a comparison with class-A and class-C β-lactamases. *Eur. J. Biochem.* **207**, 97–102

26. Wilkin, J.M., Jamin, M., Damblon, C., Zhao, G.H., Joris, B., Duez, C. & Frère, J.M. (1993) The mechanism of action of DD-peptidases: the role of tyrosine 159 in the *Streptomyces* R61 DD-peptidase. *Biochem. J.* **291**, 537–544

27. Wilkin, J.M., Jamin, M., Joris, B. & Frère, J.M. (1993) The mechanism of action of DD-peptidases: the role of asparagine 161 in the *Streptomyces* R61 DD-peptidase. *Biochem. J.* **293**, 195–201

28. Wilkin, J.M., Dubus, A., Joris, B. & Frère, J.M. (1994) The mechanism of action of DD-peptidases: the roles of threonines 299 and 301 in the *Streptomyces* R61 DD-peptidase. *Biochem. J.* **301**, 477–483

29. Gibson, R.M., Christensen, H. & Waley, S.G. (1990) Site-directed mutagenesis of β-lactamase I: single and double mutants of Glu-166 and Lys-73. *Biochem. J.* **272**, 613–619

30. Monnaie, D., Dubus, A. and Frère, J.M. (1994) The role of lysine-67 in a class C β-lactamase is mainly electrostatic. *Biochem. J.* **302**, 1–4

31. Jacob, F., Joris, B., Dideberg, O., Dusart, J., Ghuysen, J.M. & Frère, J.M. (1990) Engineering a novel β-lactamase by a single point mutation. *Protein Eng.* **4**, 79–86

32. Brannigan, J., Matagne, A., Jacob, F., Damblon, C., Joris, B., Klein, D., Spratt, B.G. & Frère, J.M. (1991) The mutation Lys-234 →His yields a class A β-lactamase with a novel pH-dependence. *Biochem. J.* **278**, 673–678

33. Lamotte-Brasseur, J., Dive, G., Dideberg, O., Charlier, P., Frère, J.M. & Ghuysen, J.M. (1991) Mechanism of acyl transfer by the class A serine β-lactamase of *Streptomyces albus* G. *Biochem. J.* **279**, 213–221

34. Strynadka, N.C.J., Adachi, H., Jensen, S.E., Johns, K., Sielecki, A., Betzel, C., Sutoh, K. & James, M.N.J. (1992) Molecular structure of the acyl-enzyme intermediate in β-lactam hydrolysis at 1.7 Å resolution. *Nature (London)* **359**, 700–705

35. Dowson, C.G., Hutchison, A., Woodford, N., Johnson, A.P., George, R.C. & Spratt, B.G. (1990) Penicillin-resistant viridans streptococci have obtained altered penicillin-binding protein genes from penicillin-resistant strains of *Streptococcus pneumoniae. Proc. Natl. Acad. Sci. U.S.A.* **87**, 5858–5862

36. Adam, M., Damblon, C., Jamin, M., Zorzi, W., Dusart, V., Galleni, M., el Kharroubi, A., Piras, G., Spratt, B.G., Keck, W., *et al.* (1991) Acyltransferase activities of the high molecular weight, essential penicillin-binding proteins. *Biochem. J.* **279**, 601–604

37. Tipper, D.J. & Strominger, J.L. (1965) Mechanism of action on penicillins: a proposal based on their structural similarity to acyl-D-alanyl-D-alanine. *Proc. Natl. Acad. Sci. U.S.A.* **54**, 1133–1140

2

Sphingolipid activator proteins

Kunihiko Suzuki

Brain and Development Research Center, Department of Neurology and Psychiatry, University of North Carolina School of Medicine, Chapel Hill, NC 27599, U.S.A.

Introduction

Sphingolipids are a group of complex lipids which contain a long-chain base, sphingosine, as the basic building block. In almost all naturally occurring sphingolipids, sphingosine is acylated by a long-chain fatty acid to form ceramide. A complex side-chain, consisting of carbohydrate, sialic acid and other constituents, is attached to the terminal OH-group of sphingosine. Combinations and permutations of the side-chain constituents give astronomical numbers of possibilities for different sphingolipids (Figure 1). Sphingolipids are important structural constituents of all cellular membranes of higher organisms and have diverse physiological functions. The complex sphingolipids undergo physiological turnover as membrane constituents within the

Figure 1. Structure of sphingosine and ceramide
Structures of some of the lipids frequently mentioned in the text are also indicated. X = substitutions: X = H, ceramide; X = β-galactose, galactosylceramide; X = β-glucose, glucosylceramide; X = β-galactose 3-sulphate, sulphatide; X = phosphorylcholine, sphingomyelin. For others, see Figure 2.

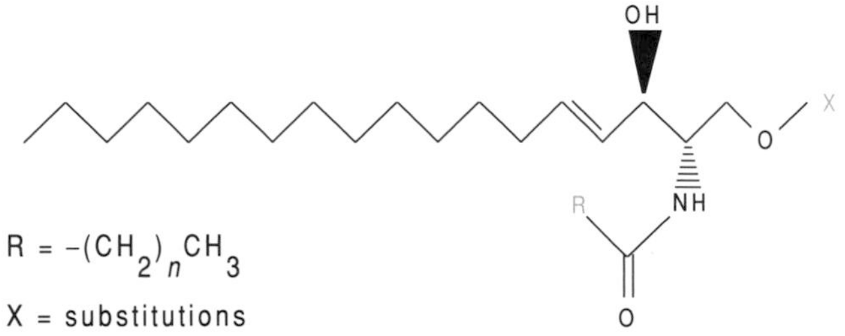

$$R = -(CH_2)_n CH_3$$

X = substitutions

lysosome by the action of a series of lysosomal hydrolytic enzymes, which sequentially remove individual residues of the side-chain, eventually to ceramide. Ceramide is finally cleaved to give sphingosine and a fatty acid. The hydrolytic enzymes are, for the most part, relatively specific for each of the degradative steps. Human genetic disorders are known to be caused by defective degradation of sphingolipids at many degradative steps, owing to genetic abnormalities of the specific enzymes for the respective steps (Figure 2).

The water-insoluble and/or membrane-embedded nature of sphingolipids presents a unique problem for the degradative enzymes, which are much more hydrophilic than their substrates. In an aqueous system, sphingolipids form either micelles or aggregates, which are inaccessible to the enzyme. When embedded in the membrane, they are also inaccessible to the enzyme, particularly when their hydrophilic side-chains are relatively short. The conventional 'trick' of measuring the activity of these enzymes against the natural sphingolipid substrates *in vitro* is to add appropriate detergents to disperse the lipids. Obviously such high concentrations of detergent do not exist in the intra-lysosomal environment where physiological degradation of these lipids takes place. Since the late 1960s, a series of small glycoproteins, which are required for hydrolysis of some of the sphingolipids *in vivo*, has been described. They are generically termed the sphingolipid activator proteins

Figure 2. Major sphingolipids in mammalian organs and their structural and metabolic relationships
Trivial names used in the text are also indicated. Sphingoglycolipids which contain sialic acid (N-acetylneuraminic acid, NeuNAc) are defined as gangliosides. Generally, these compounds are synthesized by sequential addition of sugars/sialic acids to the carbohydrate chains in the reverse direction of the arrows connecting neighbouring compounds and degraded in the direction of the arrows by their sequential removal.

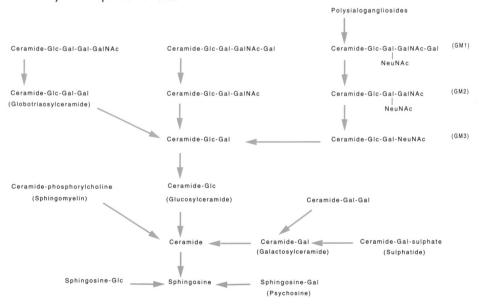

(SAPs). In recent years, many unexpected properties have been uncovered since molecular cloning of these proteins was first achieved. Simultaneously, their discovery has led to new questions being posed regarding the functions of these proteins. Five established or putative SAPs are known at this time. This article will briefly review some of the established facts, postulated — but not rigorously proven — hypotheses and potential future directions regarding the SAPs. By design, this is a selective, critical review. The requirement for a limited number of references does not allow full citation of all pertinent references. Readers are referred to some of the recent articles on this subject[1-3] for more comprehensive factual information.

Brief history

The history of the sphingolipid activator protein goes back to the mid-1960s, when Mehl and Jatzkewitz[4] observed during their attempt at purification of arylsulphatase A that their preparation lost the ability to cleave the sulphate group from its lipid substrate, sulphatide, after one particular step, and that the activity could be restored by combining the enzyme fraction with another fraction separated during that step (SAP-1). That the presence of the enzyme was detected in the fraction with a water-soluble substrate, p-nitrocatechol sulphate, in the absence of SAP-1, indicated the fundamental property of the activator protein — it is required specifically for degradation of the natural lipid substrate but not for water-soluble artificial substrates. The second sphingolipid activator protein (SAP-2) was first described by Ho and O'Brien in the tissues of patients with Gaucher disease, as a protein factor that stimulated the degradation of glucosylceramide[5]. Presence of the third sphingolipid activator protein (SAP-3) was indicated when Conzelmann and Sandhoff[6] observed that a non-enzymic protein factor was missing in a patient who exhibited the clinical, pathological and biochemical characteristics of Tay–Sachs disease without deficiency of the β-hexosaminidase α- or β-subunits. Inability of the patient's tissues to degrade GM2-ganglioside could be corrected by a protein factor extracted from normal kidney. Thus SAP-3 was the first sphingolipid activator protein discovered on the basis of its disease-causing genetic deficiency; however, equivalent genetic disorders owing to defective SAP-1 and SAP-2 were also identified later[7,8]. The cDNAs and the genes coding for the SAPs have been cloned and characterized in recent years and the disease-causing mutations identified[1-3]. When cDNA coding for the human SAP-1 was cloned, it showed that the gene encodes a large precursor protein which includes homologous sequences of both SAP-1 and SAP-2 in tandem. Furthermore, the precursor protein included two additional homologous domains flanking the SAP-1 and SAP-2 sequences. In addition to these SAPs, Vaccaro and co-workers described an apparently different protein which stimulated hydrolysis of glucosylceramide by glucosylceramidase[9]. This protein has not been as well-characterized as the other five SAPs.

Nomenclature

Over the years, the nomenclature of the sphingolipid activator proteins has become confusing. The terminology of SAP-1,2,3 was historically the first system — used by Wenger to indicate sphingolipid activator proteins, numbered in the order of discovery. Since the discovery that a single gene codes for four homologous, established, putative sphingolipid activator proteins, two different nomenclature systems have been proposed. Table 1 summarizes these nomenclature systems, together with the more traditional names for individual sphingolipid activator proteins. In the following article, the nomenclature of Fürst and Sandhoff[1] will be used primarily for the sake of consistency, supplemented by other names as appropriate. The term SAP will be used to mean all sphingolipid activator proteins as a category. Some workers express reservation regarding the term 'saposin', which implies the detergent function as the mechanism underlying their activator capacity. The mechanism of activation may well be much more complex than detergent-like solubilization of the water-insoluble substrates.

Molecular genetics

Two genes code for the five known SAPs[10–15]. The GM2-activator protein is uniquely coded by a gene located on human chromosome 5, while the other

Table 1. SAPs and their nomenclature

O'Brien et al.	Fürst and Sandhoff	Other names
GM2-activator	GM2-activator	SAP-3
Prosaposin	sap-precursor	
Saposin A	sap-A	
Saposin B	sap-B	SAP-1, sulphatide activator, GM1-activator, trihexosylceramide activator, A_2-activator protein, non-specific activator
Saposin C	sap-C	SAP-2, Gaucher factor, glucosylceramide activator, co-β-glucosidase
Saposin D	sap-D	component C

The gene coding for the GM2-activator is termed GM2-activator gene by most investigators, while the gene coding for the four other SAPs is called prosaposin by O'Brien's group and the sap-precursor by Sandhoff's group. Preprosaposin is the nascent polypeptide generated by the gene, while prosaposin is the nascent polypeptide without the signal sequence. The glucosylceramidase activator described by Vaccaro et al.[9] appears distinct from any of these activator proteins.

four homologous activators are coded by a single sap-precursor gene located on human chromosome 10 (Figures 3 and 4). There is evidence for at least one pseudogene for the GM2-activator protein. Coincidentally, the 5'- segments of both genes have so far eluded the efforts of several laboratories. Except for a very long 3'-untranslated segment, there is nothing unusual about the processed GM2-activator transcript. The open frame codes for a precursor protein of 193 amino acids, including a signal sequence of 23 amino acids. The remainder is presumably processed proteolytically at the N-terminus to give a mature GM2-activator protein of 162 amino acids. The protein is glycosylated at the only site toward the N-terminus. In contrast, the structure of the sap-precursor gene has several interesting features. The gene generates a large transcript that codes a prepro-protein of 524, 526 or 527 amino acids, depending on whether or not the very short exon 8 (9 bp) is spliced in or out during the transcript processing. At the N-terminus, 16 amino acids appear to function as the signal sequence. The remainder of the polypeptide includes four homologous domains (sap-A, sap-B, sap-C and sap-D), two of which correspond to the long-known SAP-1 (sap-B) and SAP-2 (sap-C), present in tandem in the middle part of the mRNA. Two additional homologous domains flank these two sphingolipid activators. These domains are approximately 80 amino acids

Figure 3. GM2-activator gene
The top line indicates the size of the genomic clone and a restriction map. E, *Eco*RI; B, *Bam*HI; S, *Sac*I; X, *Xba*I; K, *Kpn*I. Open boxes indicate the protein-coding sequence. mRNA is generated from the transcript by splicing at least four exons. Since the 5'-end of the gene has not been isolated, it is possible that there is still more than one exon in the upstream region. The processed mRNA we isolated is approximately 2.5 kb long and has a very long 3'-untranslated segment. Modified from Klima et al[11].

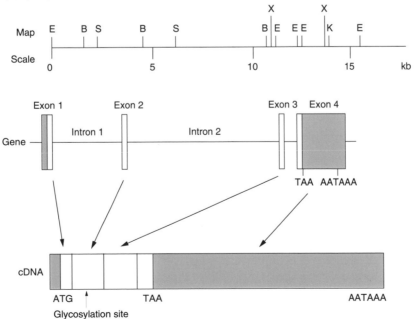

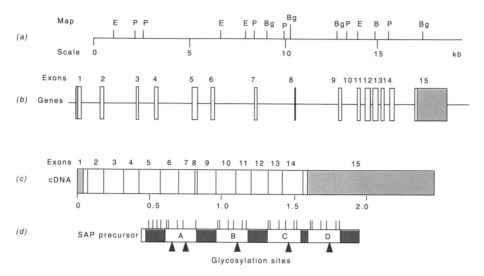

Figure 4. The sap-precursor gene

(a) The top line indicates the size of the genomic clone and a restriction map: E, *EcoRI*; P, *PstI*; Bg, *BglII*; B, *BamHI*. (b) mRNA is generated from at least 15 exons, including exon 8, which consists of only nine bases. This exon contributes zero, six or nine bases to the processed mRNA. (c) Open boxes indicate the protein-coding sequence. Since the 5'-end of the gene has not been isolated, it is possible that there is still more than one exon in the upstream region. (d) The SAP-precursor protein is proteolytically processed after translation to four homologous SAPs, indicated as A, B, C and D. One glycosylation site is conserved in all four domains. Thin vertical lines above the precursor structure indicate positions of cysteine residues. All four domains have six conserved cysteine residues. Modified from Holtschmidt *et al.*[14]

each and are 23–39% identical in amino acid sequence. Six cysteine residues and one putative glycosylation site are strictly conserved among the four domains. Only sap-A has an additional putative glycosylation site. The extra two- or three-amino acid segment derived from exon 8 is within the sap-B domain. Thus, at least from the transcript, there should be three different forms of sap-B, because the three types of mRNA are present in similar amounts.

Activator function *in vitro*

Table 2 provides an overview of reported activation of sphingolipid hydrolysis *in vitro* by the activator proteins. This is a relatively uncritical compilation of reported data in the sense that some of the assays were done in violation of the true test criteria for SAPs. The basic concept of the activator protein is, as stated above, that it stimulates hydrolysis of highly water-insoluble substrates in a detergent-free environment. Some of the reported activation used water-soluble artificial substrates. Some others tested against natural lipid substrates but in the presence of detergents. In the author's opinion, results obtained under these conditions *in vitro* cannot necessarily be extrapolated to the lipid hydrolysis in the detergent-free conditions *in vivo*. Although GM2-activator

Table 2. Specificity of SAPs *in vitro*

SAP	Activation *in vitro* (by enzyme)
GM2-activator	GM2-ganglioside (β-hexosaminidase A)
	GA2 (β-hexosaminidase A)
sap-A	Glucosylceramide (glucosylceramidase)
	Galactosylceramide (galactosylceramidase)
sap-B*	Sulphatide (arylsulphatase A)
	GM-ganglioside (acid β-galactosidase)
	Globotriaosylceramide (α-galactosidase A)
	Sphingomyelin (acid sphingomyelinase)
sap-C	Glucosylceramide (glucosylceramidase)
	Galactosylceramide (galactosylceramidase)
	Sphingomyelin (acid sphingomyelinase)
sap-D	Sphingomyelin (acid sphingomyelinase)

*In addition to the lipids listed, Li and co-workers described the stimulation of degradation of 20 natural and chemically modified lipids by six different enzymes ('non-specific activator').

protein as well as the sap-B and sap-C are known to exist in glycosylated forms, the carbohydrate chains are not necessary for the activator function, as demonstrated by deglycosylated preparations[16,17], and also with a preparation generated by a prokaryotic expression system[18]. On the other hand, a mutation which abolishes the only glycosylation site in sap-B is known to cause the clinical sap-B-deficiency disease. Even though the carbohydrate chain is not required for the activity, its absence results in a failure to route the synthesized sap-B to the lysosome because the mutant sap-B cannot acquire the mannose 6-phosphate marker needed for proper routing. In addition, the change in the primary amino acid sequence at the normal glycosylation site itself might also affect its folding/processing/routing or its activity.

Physiological activator function *in vivo* and genetic disorders

When a clinical disorder results from a genetic abnormality of a SAP, it provides firm evidence that the protein is indeed essential for normal function of the organism. As already mentioned, GM2-activator protein was first discovered as a factor in normal tissues that could restore the capacity to hydrolyse GM2-ganglioside by tissues from a patient who showed a Tay–Sachs-like phenotype without the associated enzymic defect. Two other activator proteins, sap-B and sap-C, had been known before patients were identified as having genetic defects in these proteins. In recent years, mutations in the respective genes have been demonstrated in patients suspected of having genetic SAP deficiency. Table 3 lists currently known genetic disorders caused by deficiency of the SAPs, including the known mutations so far described.

It is of the utmost importance to distinguish conceptually, as well as in practice, the physiological functions of the sphingolipid activators *in vivo* from those demonstrable in the test-tube. Activation *in vitro* does not necessarily mean physiologically essential or even significant activation *in vivo*. In this regard, the physiological essentiality of sap-B, sap-C and GM2-activator is definitive, since genetic deficiency of each of these proteins results in a clinical disease. Even in these instances, there is a large discrepancy between the observations *in vivo* and *in vitro*. Both sap-B and sap-C have been shown to activate degradation of many sphingolipids *in vitro* (Table 2); however, genetic sap-B deficiency results in a metachromatic leukodystrophy-like disorder without, for example, the phenotype associated with β-galactosidase deficiency — even though GM1-ganglioside degradation can be activated by sap-B. Similarly, sap-C deficiency results in a Gaucher-like disorder. Two interpretations are possible. The activation phenomenon demonstrated *in vitro* for the large number of substrates for sap-B could be just owing to *in vitro* artifacts and have no counterpart *in vivo*. Alternatively, there may be redundancies in the functions among the activators *in vivo*, with respect to substrates, and loss of one activator could be compensated for by other normal activators. One distinction

Table 3. Genetic disorders caused by sphingolipid activator deficiency

Genetic deficiency	Clinical phenotype	Accumulation	Mutations
GM2-activator	Tay–Sachs-like disorder Mimics β-hexosaminidase A deficiency	GM2-ganglioside and asialo GM2-ganglioside	Cys-107 → Arg Arg-169 → Pro,
All four SAPs	Only two cases from a single family known. Rapidly progressive to death at 16 weeks		Initiation codon ATG → TTG
sap-A	None known	–	–
sap-B	Metachromatic leukodystrophy-like disorder Mimics arylsulphatase A deficiency	Sulphatide and globotriaosylceramide	Thr-217 → Ile (loss of glycosylation site). 33 bp insertion after normal Gly-777. Cys-241 → Ser
sap-C	Gaucher-like disorder Mimics glucosylceramidase deficiency	Glucosylceramide	Cys-382 → Gly Cys-385 → Phe
sap-D	None known	–	–

between metachromatic leukodystrophy due to genetic arylsulphatase A deficiency and the disease caused by sap-B deficiency is excretion of globotriaosylceramide in urine in the latter, a finding suggestive of α-galactosidase activation as another physiologically essential function of sap-B. Despite the numerous other sphingolipids, degradation of which sap-B can activate *in vitro*, there is no firm evidence at this time that any of these is physiologically significant.

Physiological significance of sap-A and sap-D is less clear at this time, since demonstration of activation of sphingolipid degradation is limited to experiments *in vitro*, not always with natural lipid substrates in the absence of detergents, and since no patients with a specific defect in either of these two 'activators' are known. In this sense, sap-A and sap-D remain 'putative' SAPs to be further defined. It is, however, likely that either or both of them have physiological activator functions for degradation of some sphingolipids, because the recently discovered patients with total SAP deficiency, due to a mutation in the initiation codon of the sap precursor gene[19], showed a pattern of multiple sphingolipid abnormalities in the tissue[20], more complex than to be expected from absence of only sap-B and sap-C. On the other hand, these patients did not show any accumulation of sphingomyelin in their tissues, even though three SAPs have been reported to have activator function for sphingomyelin degradation *in vitro* (Table 2). In the author's judgment, this observation logically leads to a conclusion that none of the SAPs is required for physiological degradation of sphingomyelin *in vivo*.

Other possible functions

When the human sap-precursor gene was cloned, there were more surprises than the discovery of the four homologous sap domains encoded by a single transcript. It was quickly shown that the major sulphated glycoprotein of the Sertoli cell of the rat was similar to the human sap-precursor gene[21]. Many investigators now believe that, in fact, the Sertoli cell major glycoprotein and the four SAPs are products of the same gene. Furthermore, the sap-precursor gene has close similarities with the lung surfactant-associated protein[22]. Then it was found that some tissues contain predominantly the precursor form, rather than the processed activator proteins[23], and that the precursor form was also found in various body fluids[24]. Kishimoto, O'Brien and co-workers have shown that both individual SAPs and the precursor have the capacity to bind and transport sphingolipids. (For details of discussions on this aspect, see the recent review in reference 2.) Potier and co-workers proposed that the same gene might also code for a sialidase (α-neuraminidase), based on the observation that there are similarities between the sequences of some viral sialidases and the sap-precursor gene and on some immunological evidence. This hypothesis was greeted with scepticism, and alternative explanations have been proposed, although it has not been rigorously excluded. Recently, Horowitz and co-workers examined the sap-precursor gene expression in developing mice[25]. They found high expression in the hind brain, dorsal ganglia and genital ridge in the embryonic mouse. Sap-mRNA was not only expressed in the testis, it was also expressed in mature female gonads. These findings suggested a physiological role of the gene product(s) in the reproductive system and possibly during development.

Future investigations

In vivo versus in vitro activation

As discussed above, there is an enormous discrepancy between activation of sphingolipid degradation by the activator proteins as demonstrated in the test-tube and as can be inferred by observations in patients and other living systems. There are two divergent schools of thought among investigators active in this field. The first considers the findings in the test-tube as being directly indicative of the physiological function, while the other views in vitro data with great scepticism and considers them, in principle, to be an unreliable indicator of the physiological function. The former theory maintains, for example, that sap-C deficiency alone cannot cause the Gaucher-like disorder because sap-A is as good an activator of glucosylceramide degradation in vitro as is sap-C. The latter counters that, since patients with the Gaucher-like disorder with normal glucosylceramidase activity have been shown to have mutations in the sap-C domain, the disease is caused by the sap-C defects and only sap-C is the essential activator component for degradation of glucosylceramide in vivo — despite the observation of glucosylceramidase activation by sap-A in vitro. Even the first school of thought agrees that not all sphingolipid activation demonstrated in vitro is meaningful in vivo, in that genetic sap-B deficiency results in a metachromatic leukodystrophy-like disease with accumulation of primarily sulphatide, and possibly also globotriaosylceramide, but not GM1-ganglioside or sphingomyelin. If sap-B were responsible for degradation in vivo of all the lipids reported in vitro, its deficiency would have caused a very complex disorder affecting many lipids. Lack of sphingomyelin accumulation in the patient with total SAP deficiency, despite activation in vitro of its degradation by three SAPs, has already been commented on (see above). Further investigations should resolve these questions. In this regard, it would be critically important to adhere to the rigorous criteria for testing even activator activity in vitro. Activation should be tested with natural lipid substrates in the absence of any detergents. Activation observed in the test-tube, with either water-soluble artificial substrates or in the presence of detergents, should be viewed with great caution.

Functions of the three possible sap-B Isoforms

As mentioned, three different forms of precursor mRNA exist, each having a zero-, six- or nine-base insertion within the sap-B domain. However, when sap-B was purified from tissues and microsequenced, no sap-B containing an extra two or three amino acids was found. If they existed in significant proportions of the total, evidence of their presence should have been observed in the microsequencing data beyond the insertion point. Several possibilities need to be sorted out: all three sap-B isoforms may be functional as the sulphatide activator; only sap-B without insertion may be active as an activator and the isoforms with extra amino acids may be artifacts of Nature and unstable; or the precursor with the extra amino acids may be destined to be processed differently to generate functionally different peptides. However, the three forms of the mRNA, when expressed in the baculovirus system, generated precursor proteins which then appeared to be processed identically (T. Nishigaki, M. Henseler, M. Schröder, L. Sandhoff and K. Suzuki, unpublished work).

Possible multiple functions

Perhaps the most important question concerning the sap-precursor gene is the possible multiple function of its products. Differential distribution of the precursor and the processed SAPs in different organs, the sphingolipid activator functions, presence of sap-precursor as the major glycoprotein in Sertoli cells, functions in the reproductive system, lipid binding/transport and the possible relationship to sialidase have been commented on already. A very detailed distribution study of the sap-precursor has been described recently[26]. The authors were careful in the use of the term, 'prosaposin-like'. Most studies on the sap-precursor protein depend on either its immunochemical reactivity and/or its size. While the identity of the N-terminus has been demonstrated, the remote possibility that some of the 'precursor' may be generated by substitution of certain exons with others by alternate splicing — and thus may not in fact be the sap-precursor but another product of the same gene — has not been rigorously excluded.

Mouse model of SAP deficiencies

Many of the remaining questions concerning SAP can be answered if appropriate laboratory animal models are available that are amenable to experimental manipulations. At least two, and, more probably, multiple laboratories are actively working to generate a genetic mouse model of SAP deficiency by gene targeting/transgenic technology. It is possible that such a model will soon become available to the scientific community.

Note added in proof: recently, sap-D has been demonstrated to be a specific activator *in vivo* in cultured fibroblasts[27] as well as *in vitro*[28].

The author thanks Professor Konrad Sandhoff for making a copy of his review chapter available before publication[3]. Much of the work described in this article represents results of a close collaboration for several years between the author's laboratory and that of Professor Konrad Sandhoff, Bonn, Germany, as should be clear from cited references. However, the author is fully responsible for judgemental statements on as yet unresolved, or controversial, matters. Work from the author's own laboratory has been supported over the years by a series of research grants from the NIH, in recent years by RO1-NS-28997, RO1-NS-24289 and a Mental Retardation Research Center Core grant, P30-HD-03110. The collaboration with the Sandhoff laboratory has been facilitated by the Senior Scientist Award from the Alexander von Humboldt Stiftung to the author.

References

1. Fürst, W. & Sandhoff, K. (1992) Activator proteins and topology of lysosomal sphingolipid catabolism. *Biochim. Biophys. Acta* **1126**, 1–16

2. Kishimoto, Y., Hiraiwa, M. & O'Brien, J.S. (1992) Saposins: structure, function, distribution, and molecular genetics. *J. Lipid Res.* **33**, 1255–1267

3. Sandhoff, K., Harzer, K. & Fürst, W. (1995) Sphingolipid activator proteins, in *The Metabolic Basis of Inherited Disease* (Scriver, C.R., Beaudet, A.L., Sly, W.S. & Valle, D., eds.), McGraw-Hill, New York, in the press

4. Mehl, E. & Jatzkewitz, H. (1964) Eine Cerebrosidsulfatase aus Schweineniere. *Hoppe-Seylers Z. Physiol. Chem.* **339**, 260–276

5. Ho, M.W. & O'Brien, J.S. (1971) Gaucher's disease: deficiency of "acid" β-glucosidase and reconstitution of enzyme activity *in vitro*. *Proc. Natl. Acad. Sci. U.S.A.* **68**, 2810–2813

6. Conzelmann, E. & Sandhoff, K. (1978) AB variant of infantile GM2 gangliosidosis: deficiency of a factor necessary for stimulation of hexosaminidase A-catalyzed degradation of ganglioside GM2 and glycolipid GA2. *Proc. Natl. Acad. Sci. U.S.A.* **75**, 3979–3983

7. Stevens, R.L., Fluharty, A.L., Kihara, H., Kaback, M.M., Shapiro, L.J., Marsh, B., Sandhoff, K. & Fischer, G. (1981) Cerebroside sulfatase activator deficiency induced metachromatic leukodystrophy. *Am. J. Hum. Genet.* **33**, 900–906

8. Christomanou, H., Aignesberger, A. & Linke, R.P. (1986) Immunochemical characterization of two activator proteins stimulating enzymic sphingomyelin degradation *in vitro*: absence of one of them in a human Gaucher disease variant. *Biol. Chem. Hoppe-Seyler* **367**, 879–890

9. Vaccaro, A.M., Muscillo, M., Gallozzi, E., Salvioli, R., Tatti, M. & Suzuki, K. (1985) An endogenous activator protein in human placenta for enzymatic degradation of glucosylceramide. *Biochim. Biophys. Acta* **836**, 157–166

10. Schröder, M., Klima, H., Nakano, T., Kwon, H., Quintern, L.E., Gärtner, S., Suzuki, K. & Sandhoff, K. (1989) Isolation of a cDNA encoding the human GM2-activator protein. *FEBS Lett.* **251**, 197–200

11. Klima, H., Tanaka, A., Schnabel, D., Nakano, T., Schröder, M., Suzuki, K. & Sandhoff, K. (1991) Characterization of full-length cDNAs and the gene coding for the human GM2 activator protein. *FEBS Lett.* **289**, 260–264

12. Nakano, T., Sandhoff, K., Stümper, J., Christomanou, H. & Suzuki, K. (1989) Structure of full-length cDNA coding for sulfatide activator, a co-β-glucosidase and two other homologous proteins: two alternate forms of the sulfatide activator. *J. Biochem.* **105**, 152–154

13. O'Brien, J.S., Kretz, K.A., Dewji, N.N., Wenger, D.A., Esch, F. & Fluharty, A.L. (1988) Coding of two sphingolipid activator proteins (SAP-1 and SAP-2) by same genetic locus. *Science* **241**, 1098–1101

14. Holtschmidt, H., Sandhoff, K., Fürst, W., Kwon, H.Y., Schnabel, D. & Suzuki, K. (1991) The organization of the gene for the human cerebroside sulfate activator protein. *FEBS Lett.* **280**, 267–270

15. Rorman, E.G., Scheinker, V. & Grabowski, G.A. (1992) Structure and evolution of the human prosaposin chromosomal gene. *Genomics* **13**, 312–318

16. Sano, A. & Radin, N.S. (1988) The carbohydrate moiety of the activator protein for glucosylceramide β-glucosidase. *Biochem. Biophys. Res. Commun.* **154**, 1197–1203

17. Hiraiwa, M., Soeda, S., Martin, B.M., Fluharty, A.L., Hirabayashi, Y., O'Brien, J.S. & Kishimoto, Y. (1993) The effect of carbohydrate removal on stability and activity of saposin B. *Arch. Biochem. Biophys.* **303**, 326–331

18. Klima, H., Klein, A., van Echten, G., Schwarzmann, G., Suzuki, K. & Sandhoff, K. (1993) Overexpression of a functionally active human GM2-activator protein in *E. coli. Biochem. J.* **292**, 571–576

19. Bradová, V., Smíd, F., Ulrich-Bott, B., Roggendorf, W., Paton, B.C. & Harzer, K. (1993) Prosaposin deficiency: further characterization of the sphingolipid activator-deficient sibs. *Hum. Genet.* **92**, 143–152

20. Schnabel, D., Schröder, M., Fürst, W., Klein, A., Hurwitz, R., Zenk, T., Weber, J., Harzer, K., Paton, B., Poulos, A., *et al.* (1992) Simultaneous deficiency of sphingolipid activator proteins 1 and 2 is caused by a mutation in the initiation codon of their common gene. *J. Biol. Chem.* **267**, 3312–3315

21. Collard, M.W., Sylvester, S.R., Trsuruta, J.K. & Griswold, M.D. (1988) Biosynthesis and molecular cloning of sulfated glycoprotein-1 secreted by rat Sertoli cells: sequence similarity with the 70 kD precursor to sulfatide-GM1 activator. *Biochemistry* **27**, 4557–4564

22. Pathy, L. (1991) Homology of the precursor of pulmonary surfactant-associated protein, SP-B with prosaposin and sulfated glycoprotein. *J. Biol. Chem.* **266**, 6035–6037

23. Sano, A., Hineno, T., Mizuno, T., Kondoh, K., Ueno, S., Kakimoto, Y. & Inui, K. (1989) Sphingolipid hydrolase activator proteins and their precursors. *Biochem. Biophys. Res. Commun.* **165**, 1191–1197

24. Hineno, T., Sano, A., Kondoh, K., Ueno, S., Kakimoto, Y. & Yoshida, K. (1991) Secretion of sphingolipid hydrolase activator precursor, prosaposin. *Biochem. Biophys. Res. Commun.* **176**, 668–674

25. Sprecher-Levy, H., Orr-Urtreger, A., Lonai, P. & Horowitz, M. (1993) Murine prosaposin: Expression in the reproductive system of a gene implicated in human genetic disease. *Cell Mol. Biol.* **39**, 287–299

26. Kondoh, K., Sano, A., Kakimoto, Y., Matsuda, S. & Sakanaka, M. (1993) Distribution of prosaposin-like immunoreactivity in rat brain. *J. Comp. Neurol.* **334**, 590–602

27. Klein, A., Henseler, M., Klein, C., Suzuki, K. & Sandhoff, K. (1994) Sphingolipid activator protein D (sap-D) stimulates the lysosomal degradation of ceramide *in vivo. Biochem. Biophys. Res. Commun.* **200**, 1440–1448

28. Azuma, N., O' Brien, J.S., Moser, H.W. & Kishimoto, Y. (1994) Stimulation of acid ceramides by saposin D. *Arch. Biochem. Biophys.* **311**, 354–357

3

Oxygen toxicity, free radicals and antioxidants in human disease: biochemical implications in atherosclerosis and the problems of premature neonates

Catherine A. Rice-Evans and Vimala Gopinathan

Free Radical Research Group, UMDS Guy's Hospital, St Thomas's Street, London SE1 9RT, U.K.

Introduction

Oxygen, which is essential for life, is one of the major intermediaries in free radical mechanisms[1]. These mechanisms contribute to normal aerobic processes, for example bacterial killing and oxidase action, but uncontrolled activation can have deleterious consequences. Oxygen is toxic not because of its own reactivity, which is rather feeble, but because it is capable of undergoing a series of reductions to superoxide and hydrogen peroxide, the reactivity of which can be amplified by a variety of mediators. In this review, the toxicity of oxygen-derived free radicals will be discussed in the context of the mechanisms whereby endogenous sources and exogenous stresses contribute to tissue damage in coronary heart disease and to oxidative damage in disorders of premature neonates, respectively, and the protective effects of antioxidants.

There are several factors controlling endogenous release of free radicals during tissue injury[2]: (i) phagocyte recruitment and activation at sites of injury through the action of the membrane-bound oxidase of neutrophils, macrophages, monocytes and eosinophils, producing superoxide radicals; (ii) activation of other superoxide-producing enzymes, e.g. xanthine oxidase; (iii) disruption of mitochondrial electron transport during ischaemia, for example, allowing leakage of electrons on to oxygen during reperfusion; and (iv) activation of arachidonic acid metabolism and enzyme-mediated formation of hydroperoxides.

It is becoming well-recognized that reactive oxygen species, such as the superoxide radical ($O_2^-\cdot$) and hydrogen peroxide (H_2O_2), may be important mediators of cell injury during disease processes via the oxidation of membranes or the alteration of critical enzyme systems. Superoxide and hydrogen peroxide, the product of its dismutation, are not very reactive by themselves; their reactivity can be amplified by interaction with a variety of agents, including nitric oxide (NO), or by interaction of hydrogen peroxide with haem proteins, transition metal ions or myeloperoxidase (Figure 1).

NO released from endothelial cells on stimulation with bradykinin, acetyl choline and so on is synthesized in a variety of cells through an arginine-dependent NO synthase. It is implicated in the regulation of vascular tone and the control of blood flow through the stimulation of the relaxation of smooth muscle cells through the activation of guanylate cyclase. The reaction with superoxide radical $O_2^-\cdot + NO \rightarrow ONOO^-$ proceeds with a rate constant of 6.7×10^9 mol$^{-1}\cdot$s^{-1} with the formation of peroxynitrite. This is a more favourable reaction than the dismutation reaction for superoxide with superoxide dismutase (rate constant 1×10^8 mol$^{-1}\cdot$s^{-1}). It has been shown that NO inhibits the membrane-bound oxidase responsible for generating superoxide radicals in activated neutrophils, suggesting a limitation to the prospect that both NO and superoxide might be produced concomitantly in these particular cells. Peroxynitrite is a potent oxidant capable of peroxidizing polyunsaturated fatty acyl side-chains of lipids, oxidizing reduced protein thiol side-chains and

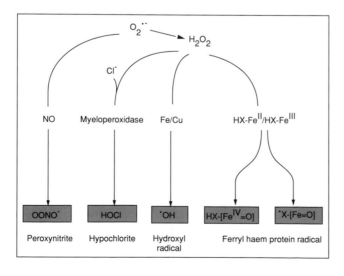

Figure 1. Amplification of the toxicity of superoxide radicals and hydrogen peroxide

inducing the nitration of tyrosine. (The biochemical, physiological and pathological aspects of NO are reviewed in reference 3.)

Phagocytic cells produce superoxide radicals as part of the body's defence mechanisms. The activation of the membrane-bound oxidase at the cell surface induces a sequence of events that produces a variety of reactive oxygen species, such as superoxide, hydrogen peroxide and hypochlorite, as shown in the equations below.

$$NADPH + 2O_2 \xrightarrow{\text{NADPH oxidase}} O_2^{-}\cdot + NADP^+ + H^+$$

$$2H^+ + 2O_2^{-}\cdot \xrightarrow{\text{superoxide dismutase}} H_2O_2 + O_2$$

$$H_2O_2 + Cl^- \xrightarrow{\text{myeloperoxidase}} HOCl + OH^-$$

Hypochlorite is cytotoxic and reacts with many biological compounds, including haem proteins, porphyrins and ascorbic acid, and many protein constituents, including thiol groups, forming disulphides and higher oxidation products, methionine forming methionine sulphoxide and amino groups forming N-chloramines.

Injury to cells and tissues may also enhance the toxicity of the active oxygen species by releasing intracellular transition metal ions into the surrounding tissue from storage sites, decompartmentalized haem proteins, or metalloproteins by interaction with delocalized proteases or oxidants. Such delocalized iron and haem proteins have the capacity to decompose peroxide to peroxyl and alkoxyl radicals which can cause further damage, including lipid peroxidation (see later), exacerbating the initial lesion.

$$LOOH + Fe^{III}\text{-complex} \rightarrow LOO\cdot + Fe^{II}\text{-complex}$$

$$LOOH + Fe^{III}\text{-complex} \rightarrow LO\cdot + [Fe^{IV}\!\!=\!\!O]\text{-complex}$$

$$LOOH + Fe^{II}\text{-complex} \rightarrow LO\cdot + Fe^{III}\text{-complex}$$

$$LO\cdot + LH \rightarrow LOH + L\cdot$$

$$L\cdot + O_2 \rightarrow LOO\cdot$$

$$LOO\cdot + LH \rightarrow LOOH + L\cdot$$

where LOOH is lipid hydroperoxide; LOO$\cdot$ is the lipid peroxyl radical; LO$\cdot$ is the lipid alkoxyl radical; and L$\cdot$ is the lipid radical. The reduction of hydrogen peroxide to hydroxyl radical ($\cdot$OH) can be mediated in the presence of such available non-protein-bound transition metals, such as iron or copper.

$$Fe^{2+} + H_2O_2 \rightarrow {}^{\cdot}OH + OH^- + Fe^{3+}$$

The hydroxyl radical produced through transition metal catalysis has often been implicated in mechanisms of tissue damage *in vivo*, but evidence for the availability of iron and copper necessary for its formation is often lacking.

The interaction of haem proteins, such as myoglobin and haemoglobin, with hydrogen peroxide leads, via a two-electron oxidation process, to the formation of ferryl haem protein radical (Figure 1). Haem proteins in the $+3$ oxidation state, e.g. met-myoglobin and met-haemoglobin, are activated to the ferryl form, which has been characterized as a radical species in which the haem iron is one oxidizing equivalent above that of met-myoglobin, forming an iron–oxoferryl complex, and one oxidizing equivalent is on the globin moiety. Haem proteins in the $+2$ state are activated to ferryl myoglobin, with two oxidizing equivalents on the iron, generating an iron (IV)–oxo complex. Oxyhaem proteins are less readily activated than the met forms to the ferryl state. This is clearly a feature of the state of oxygenation rather than the oxidation state itself, since, for example, deoxymyoglobin is several orders of magnitude more sensitive to hydrogen peroxide. This reaction may have important implications for ischaemia/reperfusion injury, since myocardial ischaemia is known to cause abrupt cellular destruction in cardiac and skeletal muscle tissues. Ferryl haem protein radicals from myoglobin and haemoglobin can react with membranes and lipoproteins.

The 4 g of iron in the human body is normally compartmented into its functional locations in the haem- and non-haem-containing and iron-binding proteins and enzymes (Figure 2). The majority (65%) of the iron is in the divalent state in haemoglobin and myoglobin, involved in the transport and storage of oxygen in erythrocytes and myocytes respectively. The rest is distributed between storage sites, predominantly in the liver, spleen and bone marrow,

Figure 2. The normal functional localizations of iron in the human body

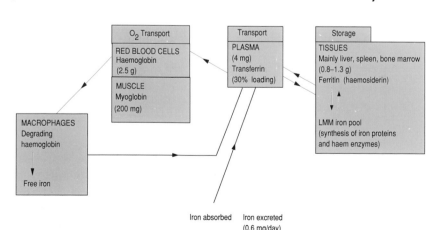

bound to ferritin, in the low-molecular-mass iron pool awaiting synthesis of iron-containing proteins and enzymes, or bound to transferrin for transport and to lactoferrin. Transferrin is normally only 30% saturated, so any iron released into plasma will immediately be mopped up in the normal course of events.

Normally the only 'free' iron available is proposed to reside in the low-molecular-mass iron pool which is normally sequestered from exerting toxic effects. Trace transition metals, or haem proteins, may be redistributed or delocalized from their normal functional locations during cell damage. In certain pathological situations, iron may be released from its normal functional compartments; mobilization of iron from ferritin by superoxide, ascorbate and other reducing agents has been demonstrated. This might occur in endothelial cells, which contain high levels of ferritin, on generation of superoxide radicals at sites of damage, as observed in the inflamed rheumatoid joint, for example, or on reperfusion post-ischaemia.

Evidence is accumulating that when haemoglobin is released via microbleeding processes, it becomes toxic; microbleeding has been observed in the eye (causing retinal damage), in the brain and at sites of inflammation. Immediately after an acute myocardial infarction, there is increased appearance of myoglobin in the blood on rupturing of cardiac myocytes. Increased local generation of hydrogen peroxide, formed from superoxide generation at the site of injury, may interact with delocalized myoglobin or haemoglobin. It has been shown in model systems *in vitro* that iron can be released from haemoglobin and myoglobin by excess hydrogen peroxide formed from superoxide radicals, and from myoglobin by lipid hydroperoxides in oxidizing low-density lipoproteins (LDLs), and this may be a potential mechanism for the availability of non-haem iron. The origin of the observed iron deposition in the artery wall in atherosclerotic lesions may be microbleeding from haemorrhaging in advanced lesions.

Antioxidants in the maintenance of health and protection against disease

Oxidative stress is a disturbance of pro-oxidant–antioxidant balance in favour of the former. Factors which can disturb the pro-oxidant–antioxidant balance are the excess production of free radicals from environmental or external sources, such as radiation (smoking, u.v. rays), hyperoxia, certain drugs and toxins, and ultrasound, or from endogenously occurring excess free radicals accompanying disease states, in which case oxidant formation is secondary to the initial lesion (reviewed in Nohl et al., 1994)[4].

Any pathological situation which increases the turnover of the antioxidant cycle, whether increased oxidative stress or modified antioxidant defences, can lead to progressive membrane, cellular and tissue damage via oxidation of lipids and proteins, affecting membrane integrity, causing altered cell permeability and leakage of intracellular components to where they should not normally be, affecting enzyme function, disrupting metalloproteins and inducing DNA damage.

In the normal course of events, cells and tissues have adequate antioxidant defences, both intracellularly and extracellularly. Of the range of antioxidants in the human body, those placed intracellularly are appropriate for dealing with aberrant generation of radicals, whereas those placed extracellularly are appropriate for binding metal ions and delocalized haem proteins and for intercepting propagating peroxidative mechanisms[5].

One of the body's first lines of defence is a system of enzymes, superoxide dismutase, glutathione peroxidase and catalase, which detoxify superoxide and hydrogen peroxide.

Enzymic antioxidant defences

(i) Superoxide dismutase disposes of superoxide radicals:

$$2O_2^{-} + 2H^{+} \rightarrow H_2O_2 + O_2$$

(ii) Catalase detoxifies hydrogen peroxide:

$$H_2O_2 \rightarrow {}^{1}/_2 O_2 + H_2O$$

(iii) Glutathione peroxidase detoxifies hydrogen peroxide and lipid peroxides with a co-substrate, glutathione:

$$2GSH + H_2O_2 \rightarrow GSSG + H_2O + {}^{1}/_2 O_2 + LOOH \rightarrow GSSG + LOH + {}^{1}/_2 O_2$$

Since certain essential minerals, including selenium, copper, manganese, zinc and iron, are components of these protective enzymes, mineral deficiencies can hamper the body's enzymic defences.

A good marker of the antioxidant status of the individual is the level, in plasma, of the range of antioxidants and free radical scavengers located therein. The antioxidant defences of human plasma include ascorbate, protein thiols, bilirubin, urate and α-tocopherol, as well as the proteins involved in iron removal — namely, caeruloplasmin and transferrin. Their actions are shown in Table 1.

The major antioxidants located in the lipophilic phase protecting the polyunsaturated fatty acid chains of the LDLs are α-tocopherol (7 mol/mol of LDL), β-carotene (1 mol/3 mol of LDL), lycopene (1 mol/5 mol of LDL), other carotenoids and ubiquinol. Interim data *in vitro* suggest a synergism between vitamins E and C.

Recent studies have pointed to the importance of the antioxidant nutrients, vitamins E and C, and β-carotene, in maintaining health, in contributing to the decreased incidence of disease and in protecting against recurrence of patho-

Table 1. Antioxidant actions in human plasma

Antioxidant	Direct removal of reactive oxygen species				Scavenging propagating reactions	Binding or interaction with metals or biological iron complexes			
	$O_2^{-\cdot}$	H_2O_2	HOCl	Singlet oxygen	LOO·	Iron	Copper	Haem proteins	Haem
Plasma soluble									
Ascorbate	+	+	+	+		Very weakly	+ (Complex still redox reactive)		
Albumin (via-SH)			+		+				+
Urate			+	+		+	+		
Bilirubin–albumin				+	+				
Bilirubin (free)				+					
Lipophilic									
α-Tocopherol				+	+				
β-Carotene				+	+ (At low oxygen tension)				
Ubiquinol					+				
Iron-controlling									
Caeruloplasmin	+					+ [$Fe^{2+} \rightarrow Fe^{3+}$]			
Transferrin						+			
Haptoglobin								+	
Haemopexin									+

logical events. More specifically, evidence suggests that the presence of the antioxidant vitamins C and E in the blood may have a protective role against cardiovascular disease.

- The World Health Organization cross-cultural epidemiological survey showed an inverse correlation between plasma α-tocopherol levels (lipid-normalized) and mortality from ischaemic heart disease. The study involved 100 middle-aged men (40–59 years old) from each of 16 different population groups: 12 out of 16 having similar blood-cholesterol and blood-pressure levels, but differing greatly in tocopherol levels and heart-disease death rates. Interestingly, the results revealed the European 'ischaemic heart disease slope', i.e. from Northern Europeans with high mortality and low levels of blood α-tocopherol to Southern Europeans with low mortality and higher plasma tocopherol levels[6].

- Studies have shown that vitamin E can help normalize an atherogenic blood lipid profile and reduce hyperlipidaemia.

- In a Scottish case–control study involving 6000 participants, the relationship between plasma concentrations of vitamins A, C and E and carotene and the risk of angina pectoris was investigated. In this study of heavily smoking males, age range 35–54 years, 110 were newly diagnosed angina sufferers compared with 394 controls. Vitamin E levels were inversely related to risk of angina after adjustment for age, smoking, blood pressure, lipids and relative weight. Related biochemical and clinical studies have shown that the relationship with β-carotene declined and that with ascorbate subsequently disappeared after adjustment for smoking.

- More recent studies reveal that it is now becoming possible to define the intake of antioxidant nutrients associated with the low subsequent incidence of disease. The Health Professionals study involving 40 000 males and 87 000 females reported a 37% and 41%, respectively, reduction in risk of coronary heart disease in those groups that had supplemented with at least 100 i.u. of vitamin E per day for more than 2 years.

- The Harvard Male Physicians study on β-carotene supplementation involved a randomized, double-blind, placebo-controlled trial of 333 male physicians, age range 40–84 years, with angina pectoris and/or coronary revascularization. The administration of 50 mg β-carotene on alternate days revealed a 44% reduction in all major coronary events defined as myocardial infarction, revascularization or death.

One of the key mechanisms relating antioxidant status in the blood and decreased risk for coronary heart disease relates to the postulation that oxidation of LDL in the artery wall is the early event in atherosclerosis. for which evidence is mounting.

Atherosclerosis: oxidants and antioxidants

Atherosclerosis is the narrowing of the arterial lumen consequent to localized thickening of the intima, and is of multi-factorial aetiology. The earliest recognizable lesion of atherosclerosis is the fatty streak, an aggregation of lipid-rich macrophages and T-lymphocytes within the intima, the innermost layer of the artery wall. There has long been an association between increased cholesterol in the blood and atherosclerosis and it is one of the major tasks of the LDLs to transport cholesterol in the blood and deliver it to cells where the LDL is recognized and taken up by the LDL receptor. Biochemical and clinical studies have suggested that oxidized LDL is atherogenic. There is increasing evidence for a role for oxidized LDL *in vivo* in the development of atherosclerotic lesions. This is excellently reviewed in Steinberg *et al.*[7] and Esterbauer *et al.*[8]

Oxidation of the polyunsaturated fatty acids in LDL may be initiated by enzymic or non-enzymic mechanisms. The former involve the action of phospholipase A_2 releasing the polyunsaturated fatty acid, with subsequent action of lipoxygenase or cyclo-oxygenase incorporating oxygen. The formation of hydroperoxides allows a mechanism of cycling through the action of redox agents, e.g. haem proteins, generating alkoxyl and peroxyl radicals which can initiate non-enzymic lipid peroxidation. Any primary free radical of sufficient reactivity may subtract an allylic hydrogen atom (Figure 3) from a reactive methylene group of polyunsaturated fatty acid side-chains. The formation of the initiating species is accompanied by bond rearrangement that results in stabilization by diene-conjugate formation. The lipid radical then takes up oxy-

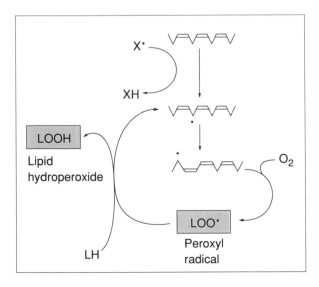

Figure 3. Mechanism of lipid peroxidation
See text for details.

gen to form the peroxyl radical. Propagation reactions follow, leading to the formation of lipid hydroperoxides. This propagation phase can be repeated many times. Thus an initial event triggering lipid peroxidation can be amplified, as long as oxygen supplies and unoxidized polyunsaturated fatty acid chains are available.

Furthermore, the accumulation of hydroperoxides and their subsequent decomposition to alkoxyl and peroxyl radicals can accelerate the chain reaction of polyunsaturated fatty acid peroxidation, leading to oxidative damage to cells, membranes and lipoproteins. It is well-recognized that haem proteins may play a role in promoting oxidative stress, through their redox-cycling properties, by catalysing the decomposition of hydroperoxides generating alkoxyl and peroxyl radical species, which can exacerbate the peroxidative process, initiating further rounds of lipid peroxidation as well as recycling the haem proteins for further oxidative events. The lipid peroxyl radicals formed during the modification of the polyunsaturated fatty acid side-chains of lipids, can amplify lipid peroxidation, oxidize cholesterol and react with proteins. Cleavage of the carbon bonds during lipid peroxidation reactions results in the formation of aldehydic products, such as cytotoxic alkanals and alkenals. The breakdown products of lipid peroxidation — alkanals, such as malonyldialdehyde, alkenals, and hydroxy alkenals, such as 4-hydroxynonenal — are toxic, undergoing Schiff base formation with amino groups and interacting with thiol groups.

Carbonyl derivatives as decomposition products of lipid peroxidation or monosaccharide oxidation can interact with amino groups on protein amino acyl side-chains, thus altering their charge and nature. This former mechanism has been proposed as a contributing sequence of events in the oxidative modification of LDLs, facilitating recognition by scavenger receptors on target macrophages. The binding of malonyldialdehyde, 4-hydroxynonenal and other secondary metabolites of the peroxidation of polyunsaturated fatty acyl chains to lysine residues on the apolipoprotein B (apoB) protein portion of LDLs decreases the positive charge on the surface of the LDL and limits its uptake by the conventional LDL receptor on cells. Rather, it becomes recognizable by the scavenger receptors on macrophages, forming cholesterol-laden foam cells, contributing towards the fatty streak in atherosclerosis.

Current theories support the idea that atherosclerosis begins with damage, by some mechanism, to the endothelium and the oxidation of LDL in the artery wall. Endothelial damage is proposed to be followed by the attachment of monocytes from the circulation, which develop into macrophages in the artery wall. Scavenger receptors on target macrophages take up oxidized LDL molecules and convert them to cholesterol-laden foam cells characteristic of the fatty streak[3] (Figure 4). Products of lipid peroxidation, such as lysophosphatidyl choline, 4-hydroxynonenal and so on, may act as chemotactic factors for blood monocytes, encouraging their recruitment into the lesioned area. Activated monocytes and macrophages may injure neighbouring endothelial

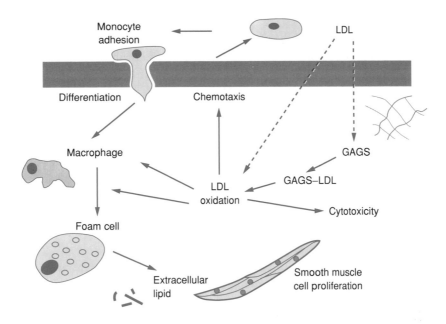

Figure 4. LDL oxidation and events in the artery wall
Reproduced from reference 3, with permission. Abbreviation used: GAGS, glycosaminoglycans.

cells by secreting superoxide radicals, hydrogen peroxide and hydrolytic enzymes, and factors released by macrophages are able to stimulate the proliferation of smooth muscle cells.

Evidence is accumulating, from studies *in vitro*, that oxidative processes might play a central role in the pathological changes involved in the pathogenesis of atherosclerosis. LDL can be oxidatively modified by a variety of systems and agents and is subsequently recognized and rapidly taken up by scavenger receptors on target macrophages. Such systems include cells in culture, including endothelial cells, arterial smooth muscle cells and monocytes; metalloproteins, such as ruptured erythrocytes and myocytes; co-incubation with transition metal ions, such as copper, or with haem proteins, such as haemoglobin and myoglobin; and interaction with haem protein-derived free radicals, with peroxynitrite, with oxygenase enzymes or by incorporation of fatty acyl hydroperoxides.

Human neutrophils have also been shown to be capable of mediating LDL oxidation such that it becomes cytotoxic, but these cells are not a common constituent of atheromatous lesions. The relative contributions of the different cell types may achieve different levels of importance at the various stages in the development of the lesion. Normal arterial wall contains endothelial cells and smooth muscle cells, whereas atherosclerotic lesions may also contain macrophages and T-lymphocytes. Thus, when the atherosclerotic lesion develops, what is the mechanism by which LDL becomes oxidized *in vivo*? Stimulation of endothelial cells, smooth muscle cells or macrophages may induce the secretion of components capable of initiating oxidation, or propagation of peroxidation may occur subsequent to lipoxygenase-mediated hydroperoxide formation. If the former, what is the probable initiating agent, where is it located and what activates it? If the latter, are the lipoxygenases macrophage-derived or from other cell sources?

Cell-induced modification of LDL *in vitro* has been demonstrated to be mediated by free radicals. One feature which all these cells have in common is the release of superoxide radicals, albeit by different mechanisms and at different rates. But it is known that superoxide radicals will not initiate the oxidation of polyunsaturated fatty acids (unless protonated). Thus, addition of superoxide dismutase, for example, has been shown to have an inhibitory effect on the oxidative modification, although the response varies according to cell type, implicating superoxide radicals in the mechanism of cell-mediated modification. The significance of superoxide radicals in the initiation, but not in the propagation, of LDL oxidation in cultures of monocytes/macrophages, is indicated by experiments showing inhibition of the oxidative modification by superoxide dismutase only if the antioxidant is added within a few hours of the initiation of the incubation. However, lipid chain-breaking antioxidants are effective in inhibiting the oxidative modification late into the oxidative process. It is important to note that small amounts of iron in the medium are an absolute requirement for oxidation of LDL by cultured macrophages, and probably for all cultured cell systems studied so far. Lipoxygenase inhibitors which are also substances with antioxidant properties are, not surprisingly, effective in inhibiting the oxidative modification induced by endothelial cells and macrophages. Thus endothelial cells can initiate the oxidation of LDL through a superoxide-independent pathway that involves lipoxygenase, and this pathway may predominate in endothelial cells. Others have reported that monocyte-mediated oxidation of LDL involves monocyte lipoxygenase products which induce release of superoxide radicals from the monocytes. Thus, of the cells of the type present in the arterial wall, activated macrophages, endothelial cells and smooth muscle cells have been reported to oxidize LDL. It is of interest to note that all these cells secrete superoxide radicals, hydrogen peroxide and hydrolytic enzymes; however, as mentioned above, the superoxide released from these cells and hydrogen peroxide generated therefrom are not, as such, very reactive towards polyunsaturated fatty acids. Their reactivity may, in principle, be amplified at lower pH values, or in the presence of available delocalized haem proteins generating ferryl haem protein-derived radicals, transition metal ions generating hydroxyl radical, or NO forming peroxynitrite (Figure 1).

Thus the question arises as to what forms of haem proteins, metalloproteins and transition metals are available *in vivo*, capable of mediating the formation of damaging initiating or propagating species. The majority of iron and haem proteins in the human body are protected *in vivo* from exerting pro-oxidant activities by their compartmentalization within their functional locations in the haem and non-haem, iron-containing proteins and enzymes (as described previously). In atherogenesis, the trapping of haem proteins released from ruptured cells in the artery wall, in the oxidizing locality of activated macrophages, may be capable of enhancing the cycling of the oxidation of LDL which has penetrated the endothelium, with the generation of further propagating species. This hypothesis presupposes the presence of available haem proteins in the sub-endothelial space. It has been demonstrated *in vitro* that not only ferryl haemoglobin and myoglobin radicals but also ruptured cardiac myocytes and erythrocytes, which generate ferryl radical species on activation, oxidatively modify LDL and induce alteration of the surface charge and recognition properties. The interaction of ruptured erythrocytes and myocytes with LDL induces oxidative damage to the LDL and the time-scale of this haem conversion is related to the antioxidant status of the LDL and that of the erythrocyte lysate. (It should be noted, however, that haemoglobin has been observed to occur freely in more advanced atherosclerotic plaques where haemorrhaging occurs.) It is conceivable that haem proteins leaking from ruptured cells may be capable of enhancing the oxidation of LDL which has penetrated the endothelium. This may occur by haem protein-mediated decomposition of preformed peroxides in LDL that has already been minimally oxidized by contact with neighbouring cells or the enzymic activity of lipoxygenases, although there is no evidence *in vivo* at present. Delocalized haem proteins may also be significant, in the appropriate location, for converting hydroperoxides into alkoxyl and peroxyl radical species, which are capable of initiating peroxidation in LDL, in the same way that transition metal complexes can propagate oxidative damage in LDL. Additionally, it has been shown in model systems *in vitro* that iron can be released from haemoglobin and myoglobin by excess hydrogen peroxide arising continually from super-oxide radicals, released from inflammatory or other superoxide-producing cells. In addition, recent studies *in vitro* have shown that LDL oxidized by haem protein-derived free radicals accumulates lipid hydroperoxides which, in a time-dependent manner, destabilize the haem ring and promote haem destruction and iron release. Any released iron may, in an appropriate environment, exert catalytic effects in the generation of other highly reactive, toxic, initiating species. Thus a source of the reported iron deposition in the artery wall in material isolated from human atherosclerotic lesions may be microbleeding from haemorrhaging in advanced lesions. In addition, recent studies have indicated the possible relationship between the risk of myocardial infarction and serum ferritin levels, and others have shown associations between plasma copper levels and the progression of atherosclerotic lesions.

Oxidation of LDL mediated by these agents is dependent on the presence of pre-formed lipid hydroperoxides or minimally modified LDL. In the presence of pre-formed lipid hydroperoxides, induced by enzymic pathways (lipoxygenase-mediated) or non-enzymic pathways (radical-mediated), propagation of peroxidation can be effected in the vicinity of haem- and iron-containing species, generating alkoxyl and peroxyl radicals which can amplify the damage by initiating further rounds of lipid peroxidation. Thus the alkoxyl radical formed is susceptible either to interaction with polyunsaturated fatty acid chains, effectively re-initiating further damage, or to interaction with a chain-breaking antioxidant, such as α-tocopherol or probucol, forming the hydroxyl fatty acyl derivative, LOH, and terminating the interaction for this species. The fate of lipid peroxyl radical passing through the same sequence of events will be lipid hydroperoxide formation, which can then re-enter the same propagative cycle catalysed by haem proteins or transition metal complexes, leading to further lipid peroxidation and oxidation, and oxidative modification of the LDL. However, this does not explain the nature and origin of the initiating species.

Esterbauer's group have studied LDL oxidation *in vitro* and have shown that there are three distinct stages in this process[8]. In the first part of the reaction, the rate of oxidation is low and this period is often referred to as the lag phase; the lag phase is apparently dependent on the endogenous antioxidant content of the LDL, the lipid hydroperoxide content of the LDL particle and the fatty acid composition. In the second or propagation phase of the reaction, the rate of oxidation is much faster and independent of the initial antioxidant status of the LDL molecule. Ultimately, the termination reactions predominate and suppress the peroxidation process.

Antioxidant status of LDL and plasma are important determinants of the susceptibility of LDL to peroxidation. Esterbauer has shown that susceptibility of LDL to oxidation is dependent on two factors: one being vitamin E-dependent and the other vitamin E-independent, but dependent on the presence of other antioxidants (the carotenoids and maybe polyphenols), the endogenous levels of preformed lipid hydroperoxides and the fatty acid composition. Of the other dietary antioxidants which may, theoretically, be localized in the LDL, the bioavailability of the flavonoids and polyphenols from red wine, tea, onions and apples is unclear as yet, although interesting studies have suggested a relationship between the increased consumption of these components and cardioprotection. Antioxidants which inhibit LDL oxidation *in vitro* prevent fatty streak formation in animal models *in vivo* and are associated with protection against coronary artery disease in population studies. In atherogenesis, trapping of haem proteins released from ruptured erythrocytes in the artery wall, in the oxidizing locality of activated macrophages, may be capable of enhancing the oxidation of LDL which has penetrated the endothelium.

A major question in considering the mechanisms relating antioxidant status in the blood with risk of coronary heart disease is the relationship between plasma antioxidant levels and the total antioxidant activity, which might depend on synergistic interactions between antioxidants. Many studies *in vitro* have discussed the synergistic interactions between vitamins E and C, and between vitamin E and β-carotene, and shown that the water-soluble vitamin C acts to keep vitamin E, located in a lipophilic region, in its reduced state. In addition, based on the standard one-electron reduction potentials, a hierarchy of antioxidants has been predicted — from highly oxidizing to highly reducing species — in terms of the ability of each reducing species to donate an electron (or hydrogen atom) to any oxidized species listed above. Thus, in general, in a cascade of free radical reactions, each reaction in the sequence will generate radicals which are less reactive, with antioxidants producing the least reactive radicals of all. However, in the presence of transition metals, such as iron or copper, and oxygen this would not be the case, but rather the reactivity would be amplified. The relative importance of the endogenous antioxidants within the LDL molecule in protecting it from oxidative modification has been demonstrated through measurement of the oxidation resistance of different LDL samples[8]. During the lag phase, the antioxidants in LDL (vitamin E, carotenoids and ubiquinol-10) are consumed in a distinct sequence: α-tocopherol is the first, followed by γ-tocopherol, and thereafter the carotenoids — cryptoxanthine, lycopene and finally β-carotene. α-Tocopherol is the most predominant antioxidant of LDL (6.4 ± 1.8 mol/mol of LDL), whereas the concentrations of the γ-tocopherol and the carotenoids, β-carotene, lycopene, cryptoxanthine, zeaxantine, luteine and phytofluene are only 0.1–0.033 that of α-tocopherol. Since the tocopherols reside in the outer layer of the LDL molecule, protecting the monolayer of phospholipids, and the carotenoids are in the inner core, protecting the cholesterol, and the progression of oxidation is likely to occur from the aqueous interface inwards, it seems reasonable to assign to α-tocopherol the rank of 'front-line' antioxidant.

There is now strong evidence that LDL oxidation does indeed occur *in vivo* and strong clinical validation of the oxidation hypothesis has been achieved. The mechanism of oxidative modification of LDL in the artery wall *in vivo* remains uncertain, but many studies have indicated that LDL extracted from atherosclerotic lesions is in an oxidized state: LDL extracted from human or animal atherosclerotic lesions has been shown to be taken up much faster than plasma LDL by macrophages by means of their scavenger receptors; antibodies that recognize oxidized LDL, but not native LDL, show positive reactivity in human or animal atherosclerotic lesions but not the normal arterial wall; auto-antibodies against oxidized LDL have been demonstrated in the plasma of patients and hypercholesterolaemic animals; antioxidants that inhibit LDL oxidation *in vitro* prevent fatty streak formation in animal models and others are associated with protection against coronary artery disease in population studies[9–11].

More information, however, is needed concerning the mechanisms involved in the oxidative modification in the artery wall *in vivo*. How is LDL oxidation initiated? How is it amplified? Would supplementation with antioxidants be the major answer towards protection? Knowledge of the precise mechanism by which oxidation is mediated is required to elucidate the precise targets for antioxidant strategies. But it must be emphasized that oxidized LDL has other properties that make it more atherogenic than native LDL: i.e.

its chemotactic action towards monocytes, its cytoxicity, its ability to stimulate the release of growth factors and cytokines, and its reported interference with the response of the arteries to NO. Might antioxidants protect against these effects, or is their major role to protect against the oxidation itself?

Several studies have asked the question that, if oxidized LDL is an important factor in atherogenesis, would its circulating concentration be expected to be higher in patients suffering from coronary artery disease? Recent studies from Sweden have confirmed that the susceptibility of LDL to oxidation is associated with the severity of coronary atherosclerosis. In middle-aged male patients with coronary heart disease, plasma contains an increased content of oxidized cholesterol as well as displaying an increased susceptibility of the LDL to oxidation *in vitro*. Our recent studies on patients with carotid and femoral artery atherosclerosis have indicated that LDL isolated from patients is already partially oxidized and that a relationship exists between the susceptibility to oxidation of the LDL and the progression of the disease.

LDL oxidation is expected to occur to a significant extent only in the arterial wall. Partly oxidized LDLs with no apolipoprotein modification may re-enter the circulation after oxidation in the artery wall or interstitial fluid of other tissues. Such partially oxidized LDLs will not be readily removed because they would not be recognized by the scavenger receptors on the cells in the liver. However, LDLs oxidized to the extent of producing oxidatively modified protein would be readily removed from the circulation and would, therefore, not be found in the plasma.

But what about the importance of the antioxidant nutrients in the maintenance of health? Many human studies on antioxidant nutrient levels — including studies on plasma levels, cross-cultural comparisons, case–control designs and prospective designs — have consistently revealed a higher risk of coronary heart disease (and cancer) at the levels shown below, versus minimal risk[9] (see Table 2).

If free radicals are involved in the multi-factorial pathophysiology of major diseases, an optimum status of essential antioxidants should reduce their risk and thus be a prerequisite of optimal health.

Table 2. Plasma levels of antioxidant nutrients versus cancer and coronary heart disease

	Higher risk	Minimal risk
Vitamin E (lipid standardized)	<25 μM	>28–30 μM
Vitamin C	<25–30 μM	>40–60 μM
β-Carotene	<0.2–0.3 μM	>0.3–0.4 μM

Hyperoxia and oxygen toxicity: oxygen radical injury in the newborn

Oxygen radical injury is a contributory pathogenic mechanism in several neonatal diseases[12]. Although the continuous improvement of neonatal care has dramatically increased the survival rate of very low birth weight (VLBW) infants, many of these infants remain chronically dependent on respiratory support for many weeks or even months. As a consequence, the incidence of complications of neonatal intensive care, such as bronchopulmonary dysplasia (BPD), retinopathy of prematurity (ROP), necrotizing enterocolitis (NEC) and intraventricular haemorrhage (IVH), have also increased and remain a major cause of mortality and morbidity in these VLBW premature infants.

Tissue damage induced by free radicals generated by hypoxic stress followed by hyperoxia, and associated with a reduced or inadequate antioxidant capacity in the preterm infants, has been proposed as a component in the development of all these conditions. The production of excess amounts of reactive oxygen species, such as superoxide radicals and hydrogen peroxide, has been implicated in the pathogenesis of oxygen toxicity in the newborn infant. This occurs during exposure to high ambient oxygen concentrations. Newborn infants, especially premature infants, have limited antioxidant protective capacity; in particular, they have been shown to be deficient in vitamin E, β-carotene, caeruloplasmin and α-antiproteinase, mainly because these components do not cross the placenta until the third trimester of gestation[13]. In addition, deficiencies in the trace metals selenium, copper and zinc, essential components of the antioxidant enzymes glutathione peroxidase and superoxide dismutase, have been reported.

The excess free radical production in BPD is attributed to hyperoxia, whereas in the cases of ROP, NEC and IVH it is proposed that hypoxia followed by periods of reoxygenation is the more likely mechanism for excess free radical production.

Infants with BPD or chronic lung disease (CLD) present with delayed resolution of respiratory distress syndrome (Figure 5), requiring supplemental oxygen and frequently intermittent positive pressure ventilation after 1 month of age. BPD or CLD of prematurity is now diagnosed if infants remain dependent on respiratory support beyond 1 month in association with an abnormal chest radiograph[14]. The incidence of BPD among ventilated infants varies, depending on gestational age, from 4.2% to as high as 50%[15]. Although the aetiology of BPD is multifactorial, oxygen toxicity may have a large part to play in the subsequent development of CLD.

ROP is a vasoproliferative retinopathy affecting predominantly the extremely low birth weight infants, weighing less than 1000 g at birth. Despite numerous human and animal studies, the aetiology of ROP is still ill-understood and prevention of this serious disease remains a major challenge. The acute proliferative form of ROP is common and in the majority of cases

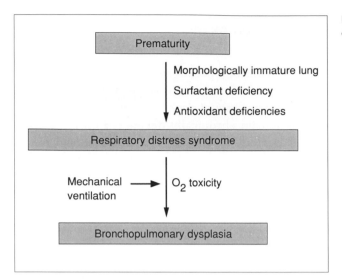

Figure 5. Pathogenesis of BPD

retinopathy regresses spontaneously; however, in some cases, ROP progresses to cicatricial sequelae which may include blindness resulting from retinal detachment.

Of infants with birth weight under 1000 g, 70–80% show acute changes, whereas above 1500 g birth weight the frequency falls to less than 10%[16]. The incidence of ROP is inversely related to birth weight and low gestation. Prolonged oxygen therapy is still the variable most closely associated with the development of ROP. After oxygen was established as an aetiological factor, subsequent restriction of oxygen usage resulted in a decrease in the incidence of ROP but this was associated with an increase in neonatal mortality and neurological morbidity in preterm infants secondary to hypoxia. Although careful control of oxygen administration reduces the incidence of ROP, significant ROP is still seen, especially in VLBW infants.

Both hypoxia and hyperoxia are known to produce similar vasoproliferative changes. In addition, oxygen may have a direct damaging effect on the developing retinal endothelium through the formation of transient free radicals to which the immature retina may be vulnerable. It has been proposed that the initial morphological event, which occurs as early as 4 days after birth, is an oxygen-induced increase in gap junctions between spindle cells which are the precursors of vascular endothelial cells[17]. It may be that the increase in gap junctions halts normal retinal vessel growth and that this may be prevented or reduced by the lipophilic antioxidant α-tocopherol[18]. On return to air breathing, the retinal vessels that have not been occluded continue to proliferate rapidly.

NEC (Figure 6) is primarily a disease of the preterm infant of VLBW (<1500 g). It occurs most frequently within the first week after birth. NEC appears to be a pathological response of the immature intestine to injury by a variety of factors. Although there is no unifying concept to the aetiology of NEC, underperfusion and/or hypoxia of the gut leading to ischaemic damage

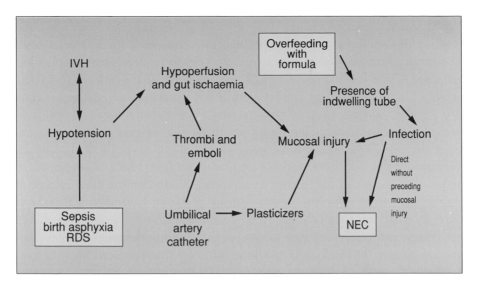

Figure 6. Aetiological factors in NEC
Abbreviation used: RDS, respiratory distress syndrome.

are thought to be among the more important aetiological factors. Aberrant patterns of blood flow in the fetal aorta, suggestive of peripheral hypoxia, may be associated with its development. The redistribution of the blood to the more important organs, such as the head, myocardium and adrenals, may predispose the neonate to NEC by sacrificing flow in the descending aorta and consequently in the splanchnic circulation[19].

IVH is the most common variety of serious intracranial haemorrhage and occurs almost exclusively in infants who are born prematurely. Because the incidence of IVH correlates directly with the degree of prematurity, and because the survival rates of very premature infants (<1000 g) continue to improve, IVH remains a major concern. Bleeding originates from the fragile vessels within the sub-ependymal germinal matrix. In approximately 80% of cases of germinal matrix haemorrhage there is spread of blood into the ventricular system (IVH). Ventricular dilatation and hydrocephalus occur commonly following moderate and severe IVH. Serial cranial ultrasonography has demonstrated that approximately 50% of IVH cases occur within the first day of life, and 90% occur up to 4 days of age[20]. IVH remains one of the most important causes of mortality and morbidity in preterm infants. The most critical determinant of long-term outcome is the extent of concomitant parenchymal injury. It must be recognized that a major IVH occurs frequently, together with a hypoxic–ischaemic cerebral insult, and the long-term neurological outcome may relate, in large part, to the concomitant or preceding hypoxic–ischaemic cerebral injury, which may be non-haemorrhagic. The two major determinants of long-term outcome after IVH are destruction of periventricular white matter by haemorrhagic infarction and post-haemorrhagic hydrocephalus.

The early signs of oxidative lung injury[21] are characterized by interstitial swelling of endothelial cells followed by destruction of Type I alveolar cells. The speed of progression of the changes appears to be dependent on the oxygen dose. Continued exposure to high oxygen concentrations leads to hyperplasia of Type II alveolar epithelial cells with marked interstitial oedema. Bronchiolar changes consist of mucosal necrosis with metaplasia and proliferation of bronchiolar epithelium with peribronchial oedema and obstruction. Over the past 20 years, much has been learned about oxygen toxicity and the mechanisms that cells have available for protection from levels of oxygen much higher than those that are usually encountered. Studies on the development of antioxidant enzymes in human lung during gestation have shown that, contrary to what was formerly believed, superoxide dismutase and glutathione peroxidase are evenly expressed throughout gestation, whereas the developmental pattern of catalase in the lung showed that the activity increases markedly throughout the gestation[22].

In an attempt to reduce oxygen-related damage, a number of agents have been administered. As mentioned previously, preterm infants are deficient in vitamin E, the chain-breaking antioxidant, since vitamin E does not cross the placenta until the third trimester of pregnancy. Although results of a non-controlled study were compatible with vitamin E reducing the incidence of the BPD, this finding has not been confirmed in randomized controlled trials.

Indirect evidence suggests that oxygen-derived free radicals, especially the superoxide radical and its more reactive products, contribute to destruction of lung tissue in the neonate. At normal oxygen tension the superoxide anion is removed by superoxide dismutase. When tissues are exposed to increased oxygen tension typically used in the treatment of neonatal lung disease this enzyme system may be overwhelmed. Superoxide anions accumulate, leading to a series of events culminating in cell membrane destruction and damage to pulmonary alveolar macrophages with release of chemotactic substances. The chemoattractants cause an outpouring of polymorphonuclear leukocytes which carry collagenases and elastase that may cause further connective tissue damage to the lung and initiate an exaggerated, abnormal repair process[15]. Investigations of the effectiveness of superoxide dismutase, in scavenging superoxide radicals, towards the prevention of BPD in a prospective double-blind controlled study have shown that this antioxidant enzyme, when given parenterally, resulted in reduced clinical and radiological evidence of BPD. These observations suggested that superoxide dismutase may be helpful in reducing the severity of BPD[23]. But this has been criticized regarding the diagnostic criteria used for BPD, and also because it is unlikely that superoxide dismutase administered in this way has access intracellularly to lung cells[24].

Reperfusion injury is initiated by biochemical events occurring during hypoxia which result in the generation of reactive oxygen metabolites, such as superoxide anion, hydrogen peroxide (and possibly hypochlorous acid).

During ischaemia, the reduced oxygen supply decreases the production of ATP which is degraded to hypoxanthine. Normally, hypoxanthine may be a substrate for xanthine dehydrogenase. During ischaemia reperfusion processes, xanthine dehydrogenase is converted to xanthine oxidase. This enzymic conversion is central to the hypothesis of oxygen radical-mediated reperfusion injury. During ischaemia, excess levels of hypoxanthine accumulate in tissues. When oxygen is reintroduced, xanthine oxidase converts hypoxanthine to xanthine with the generation of superoxide anion and hydrogen peroxide. Thus this burst of superoxide radical production initiates a cascade of events which releases reactive oxygen species and hydrogen peroxide within endothelial cells[12].

There is considerable evidence to suggest that hypoxia/reoxygenation-induced free radical formation may play an important role in the development of ROP, NEC and IVH, as in all these conditions there is injury to the microvasculature. This leads to the important question as to whether antioxidant supplementation at birth might contribute to the attenuation of these complications. The questions are, which antioxidants, at which site, using which mode of administration and targeted against which radical species? There have been several studies looking at the effect of vitamin E in preventing ROP and IVH, as mentioned previously. To date there has been little conclusive evidence.

Plasma antioxidant status in neonates

As described previously, of the range of antioxidants in the human body, those placed extracellularly are more appropriate for intercepting propagating peroxidative mechanisms and for binding metal ions and delocalized haem proteins. The antioxidant defences of human plasma include ascorbate, protein thiols, bilirubin, urate and α-tocopherol, as well as the proteins involved in iron removal: namely caeruloplasmin and transferrin.

Extremely low birth weight (<1000 g, <28 weeks gestation) premature infants have been known for many years to have limited antioxidant protective capacity. When, shortly after birth, premature infants require prolonged mechanical ventilatory support with high levels of oxygen therapy, the barotrauma, in combination with immature development of the lung antioxidant enzymes systems, contributes towards the development of BPD or CLD of the newborn.

An interesting correlation ($r^2 = 0.43$) between birth weight of premature and term neonates and total plasma antioxidant activity, as measured by the total collective response of all the antioxidants in plasma, has been noted[25] (Figure 7). Levels of the major antioxidants in the plasma of VLBW premature neonates (27 ± 2 weeks) compared with term neonates (37–41 weeks) (Table 3) show no significant differences with respect to bilirubin and α-tocopherol at birth but, as expected, the latter is very much lower than the adult range. In contrast, the levels of albumin and urate are significantly lower in the samples

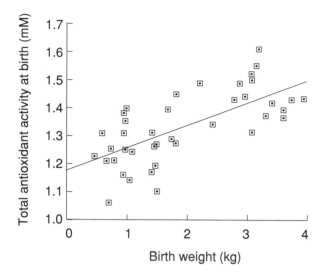

Figure 7. Relationship between total anti-oxidant activity (TAA) at birth, and the birth weight of premature and term neonates r^2 = 0.43.

from the premature babies, although the urate levels in both groups are within the normal reference interval; this only applies to the term babies for albumin.

Interestingly, elevated plasma ascorbate levels in the preterm babies at birth have been revealed, compared with term[25] (whose levels are beyond the upper limit of the adult reference interval), and yet the total plasma antioxidant activities, the response of all the collective antioxidants in plasma, is significantly lower in the preterm. This may be a marker of the increased prematurity of these neonates and might be partly a compensatory mechanism for the relatively lower proportions of urate, although not adequate enough to compensate for the lowered total antioxidant activity.

At 5 days post-partum, while the albumin levels are not changed significantly in either group, the urate levels have significantly declined in the term babies but are still within the normal range. The ascorbate levels in both groups decrease to values close to the midpoint of the normal adult range and the α-tocopherol levels progressively increase as expected. Hyper-bilirubinaemia is apparent in both groups at day 5. The mean value of total antioxidant activity has risen to values not significantly different from those for the term babies.

To assess the relative importance of the plasma antioxidant profile in relation to the total antioxidant activity, the concentration of each of the antioxidants measured has been related to its overall antioxidant potential and expressed as relative proportions to the sum of the contributions from albumin, urate, ascorbate, α-tocopherol and bilirubin (the major radical scavenging antioxidants in plasma). The results (Table 4) show that for the preterm and term babies, the combined relative contributions of urate, ascorbate and bilirubin at birth are the same, but the higher relative ascorbate and lower relative urate in the former are pronounced. When the antioxidant concentrations are expressed as relative levels in relation to the major antioxidants of human plasma, the contribution of bilirubin to the antioxidant potential at day 5 for both the term and the preterm infants is further emphasized.

Table 3. Levels of the antioxidants in the plasma of VLBW and term infants

| | Values at birth | | Values at day 5 | | |
	Premature	Term	Premature	Term	Normal adult range
Albumin	380 ± 83	482 ± 64	370 ± 80	530 ± 49	535–760
	[22]	[14]	[19]	[2]	Midpoint 640
Uric acid	267 ± 107	413 ± 103	196 ± 69	191 ± 84	180–420
	[24]	[15]	[27]	[14]	Midpoint 300
Ascorbate	164 ± 60	123 ± 28	65 ± 25	84 ± 18	34–111
	[27]	[16]	[28]	[16]	Midpoint 73
α-Tocopherol	6.8 ± 4.4	4.4 ± 1.4	10 ± 5.6	13.7 ± 4	14–44
	[10]	[10]	[23]	[8]	Midpoint 29
Bilirubin	23 ± 10	20 ± 10	167 ± 47	126 ± 92	<20
	[26]	[16]	[23]	[16]	Midpoint 10

Table 4. Relative proportions of plasma antioxidants in VLBW and term infants expressed as a percentage of the total antioxidants

	Albumin	Urate	Ascorbate	α-Tocopherol	Bilirubin
TEAC value	0.63	1.02	0.99	0.97	1.5
Reference level*	49	37.5	9.5	3	1
Term					
At birth	34.5	47.8	13.8	0.5	3.4
At day 5	41	24.1	10.2	1.6	23
Premature					
At birth	33.5	38.1	22.7	0.9	4.8
At day 5	30.8	26.4	8.5	1.2	33.1

*Reference levels taken from Miller, N.J., Rice-Evans, C., Davies, M.J., Gopinathan, V. and Milner, A. (1993) Clin. Sci. **84**, 407–412.

It is significant to note here that for the term babies a clear correlation is demonstrable (Figure 8) between the total antioxidant activity and the bilirubin levels at day 5 ($r^2 = 0.774$), showing that the bilirubin concentration has a relatively greater impact on the total antioxidant activity. This demonstrates that modestly elevated bilirubin levels may be positively favourable to infants under oxidative stress[25]. For the premature babies, the correlation is apparently much less significant ($r^2 = 0.30$), but this may be confounded by the photo-therapy treatment for hyperbilirubinaemia (and concerns of kernicterus), which would influence the bilirubin levels.

The comparative data of relative levels of antioxidants suggest that the overall deficiency in the total plasma antioxidant capacities of the preterm compared with term babies may stem from the contributions from the 'masked' oxidative stress factors, which might include the iron status and the iron-binding proteins, or phagocytic activity. In the context of the iron status, several studies have demonstrated the pressure of 'available' iron in the plasma

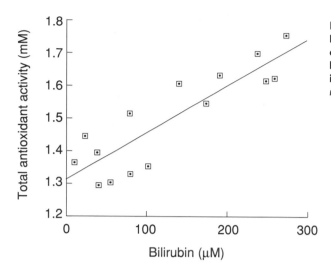

Figure 8. The correlation between total antioxidant activity and plasma bilirubin levels in term infants at day 5
$r^2 = 0.774$.

of preterm and term neonates, although the former will be more pronounced owing to several factors, including transfusion. It has been proposed by others that iron-induced oxidative stress may play a role in the pathogenesis of oxygen radical diseases of prematurity. Much work has also focused on the iron status of the newborn, and the importance of transferrin and caeruloplasmin levels in the context of antioxidant activity. However, exchange transfusion has been shown to have no influence on the total antioxidant ability of plasma in a group of neonates with rhesus haemolytic disease, despite decreases in vitamin C and bilirubin, even though iron and ferritin levels declined and the caeruloplasmin and transferrin levels and the iron-binding capacity of the plasma increased.

In summary, as the survival rates of VLBW premature infants continue to improve, oxidative injury to the tissues of these premature infants continues to be on the increase and there may be a case for studying the effects of multiple antioxidant therapy in these infants.

Catherine Rice-Evans acknowledges the British Heart Foundation, St Thomas's Trustees, the Ministry of Agriculture, Fisheries and Food, Unilever, Roche, Bioxytech and the Biotechnology and Biological Sciences Research Council for supporting the research of the Free Radical Research Group at UMDS Guy's Hospital.

References

1. Halliwell, B. & Guttridge, J.M.C. (1989) *Free Radicals in Biology and Medicine*, Clarendon Press, Oxford
2. Rice-Evans, C. (1994) Formation of free radicals and mechanisms of action in normal biochemical processes and pathological states, in *Free Radical Damage and its Control* (Rice-Evans, C. & Burdon, R.H., eds.), pp. 131–153, Elsevier, Amsterdam
3. Rice-Evans, C. & Bruckdorfer, K.R. (1992) Free radicals, lipoproteins and cardiovascular dysfunction. *Mol. Asp. Med.* **13**, 1–111

4. Nohl, H., Esterbauer, H. & Rice-Evans, C. (eds.) (1994) *Free Radicals in the Environment, Medicine and Toxicology,* Richelieu Press, London

5. Halliwell, B. (1990) How to characterise a biological antioxidant. *Free Radical Res. Commun.* **9,** 1–32

6. Gey, K.F., Puska, P., Jordan, P. & Moser, U.K. (1991) Inverse correlation between plasma vitamin E and mortality from ischaemic heart disease in cross cultural epidemiology. *Am. J. Clin. Nutr.* **53,** 326–334S

7. Steinberg, D., Parthasarathy, S., Carew, T.E., Khoo, J.C. & Witztum, J.L. (1989) Beyond cholesterol — modifications of low density lipoproteins that increase its atherogenicity. *N. Engl. J. Med.* **320,** 915–924

8. Esterbauer, H., Gebicki, J., Puhl, H. & Jurgens, G. (1992) The role of lipid peroxidation and antioxidant in oxidative modification of LDL. *Free Radical Biol. Med.* **13,** 341–390

9. Gey, F. (1994) The relationship of antioxidant status and the risk of cancer and cardiovascular disease: a critical evaluation of observational data, in *Free Radicals in the Environment: Medicine and Toxicology* (Nohl, H., Esterbauer, H. & Rice-Evans, C., eds.), pp. 181–219, Richelieu Press, London

10. Rimm, E.B., Stampfer, M.J., Ascherio, J.A., Giovanucci, E., Colditz, G.A. & Willett, W.C. (1993) Vitamin E consumption and the risk of coronary heart disease in men. *N. Engl. J. Med.* **328,** 1450–1456

11. Stampfer, M.J., Hennkens, C.H., Manson, J.E., Coldtiz, G.A., Rosner, B. & Willett, W.C. (1993) Vitamin E consumption and the risk of coronary heart disease in women. *N. Engl. J. Med.* **328,** 1444–1449

12. Saugstad, O.D. (1988) Hypoxanthine as an indicator of hypoxia: its role in health and disease through free radical production. *Paediatr. Res.* **23,** 143–150

13. Frank, L. & Sosenko, R.S. (1987) Development of lung anti-oxidant enzyme system in late gestation: possible implications for the prematurely born infant. *J. Paediatr.* **110,** 9–14

14. Northway, W.H., Jr., Rosal, R.C. & Porter, D.Y. (1987) Pulmonary Disease following respirator therapy of hyaline membrane disease: BPD. *N. Engl. J. Med.* **276,** 357–368

15. Frank, L. (1992) Antioxidants, nutrition and bronchopulmonary dysplasia. *Clin. Perinatol.* **19,** 541–562

16. Ng, Y.K., Fielder, A.R., Shaw, D.E. & Levene, M.E. (1988) Epidemiology of ROP *Lancet* **ii,** 1235–1238

17. Kretzer, F.L., Mehta, R.S., Johnson, A.T., Hunton Brown, E.S. & Hittner, H.M. (1984) Vitamin E protects against ROP through action on spindle cells. *Nature (London)* **309,** 793–795

18. Hittner, H.M., Godio, L.B. & Speer, M.E. (1983) Retrolental fibroplasia: further clinical evidence and ultrastructural support for efficacy of vitamin E in the preterm infant. *Paediatrics* **71,** 423–432

19. Malcolm, G., Ellwood, D., Devonald, K., Bulby, R., Henderson, D. & Smart, ? (1991) Absent or reversed end diastolic flow velocity in the umbilical artery and necrotizing enterocolitis. *Arch. Dis. Child.* **66,** 805–807

20. Volpe, J.J. (1989) IVH in the premature infant current concepts Part 1. *Ann. Neurol.* **25,** 3–11

21. Bancalari, E. (1988) Pathogenesis of bronchopulmonary dysplasia: an overview. *Bronchopulmon. Dysplasia Rel. Chronic Respir. Disord.* **4,** 3–15

22. McElroy, M.C., Postle, A.D. & Kelly, F.J. (1992) Catalase, superoxide dismutase and glutathione peroxidase activities of lung and liver during human development. *Biochim. Biophys. Acta* **1117,** 153–158

23. Rosenfeld, W., Evans, H., Concepcion, L., Jhaveri, R., Schaeffer, H. & Friedman, A. (1984) Prevention of BPD by administration of bovine SOD in preterm infants with respiratory distress syndrome. *J. Paediatr.* **105,** 781–785

24. Huber, N., Saifer, M.G.P. & Williams, L.D. (1980) SOD pharmacology and orgotein efficacy: new perspectives, in *Biological and Clinical Aspects of Superoxide and Superoxide Dismutase* (Bannister, W.H. & Bannister, J.V., eds.), p. 395, Elsevier, Amsterdam

25. Gopinathan, V., Miller, N.J., Milner, A.D. & Rice-Evans, C.A. (1994) Bilirubin and ascorbate: antioxidant activity in neonatal plasma. *FEBS Lett.,* **349,** 197–200

4

Reconstructed human skin: transplant, graft or biological dressing?

Edward J. Wood and Ian R. Harris

Department of Biochemistry and Molecular Biology, University of Leeds, Leeds LS2 9JT, U.K.

Introduction

The skin of a human accounts for about one-tenth of the bulk of the body, and the loss of even a part of this organ has very serious consequences. Vital functions of skin include keeping infectious and noxious agents out and keeping water vapour in. Consequently, upon serious injury, it is extremely important to restore this barrier function as soon as possible[1].

Although skin has rather remarkable regenerative properties, there is a limit to this ability. The loss of even a few square centimetres of full-thickness skin usually requires a grafting procedure (e.g. autografting from some other site on the body) if wound contracture and disfiguring scarring are not to occur. How such autografting can be carried out successfully is explained in the next section, but it is obvious that any large loss of skin area in an individual poses problems because of the limited area of donor sites available. Where someone has suffered greater than 50% loss of skin as a result of a burn injury, for example, there is a major, life-threatening problem (Figure 1). In the case of kidney or heart failure, a transplant from a relation (in the former case) or from a recently deceased person (in either case), along with the appropriate immunosuppressive therapy, can be highly successful. Such a transplant in the case of skin loss is not possible for a number of reasons and other strategies have to be employed. However, in recent years a better understanding of the biochemistry, cell biology and immunology of the human skin has made transplantation with reconstructed skin an attractive possibility.

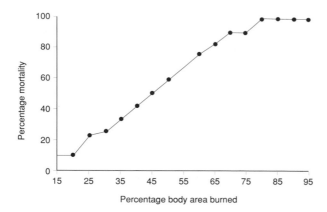

Figure 1. Mortality curve
Relation between area of body burned and chances of survival in the 45–49-year-old age range (based on statistics). Younger individuals have a higher expected survival, older individuals a lower one. Data taken from Bull, J.P. (1971) *Lancet* **2**, 1133–1134.

This article aims to reveal the background behind this potential and to show that, compared with even 10 years ago, our understanding of skin biochemistry has advanced by leaps and bounds.

The structure of human skin: regeneration and grafting

Three layers can be distinguished in skin (Figure 2)[2]. The outer epidermis is a mere 0.1 mm in thickness, but is sufficient to provide the vital barrier function mentioned above. Epidermis is totally cellular, being essentially devoid of blood vessels, connective tissue and other accessories. The major cell type present is the keratinocyte, forming a constantly renewing, keratinizing epithelium; however, in addition, there are also a few other cell types present, constituting a small percentage of the cells. These include the melanocytes, which are concerned with pigmentation; the Langerhans cells, which are immunological cells; and the Merkel cells, which are components of the nervous system. The importance of the self-renewing and regenerative properties of the epidermis is discussed below.

Beneath the epidermis is a much thicker layer, called the dermis, which constitutes the bulk of the skin. The upper layer of the dermis is based on an extracellular matrix of collagen, elastin and glycosaminoglycans containing, relatively speaking, few cells (fibroblasts) and also blood and lymph vessels, nerve endings and other structures. The third layer, beneath the dermis proper, is the hypodermis which contains a considerable amount of adipose tissue. It provides mechanical and thermal insulating properties to the skin.

The dermis has some regenerative powers, but the process of contraction of a wound is probably more important biologically to survival than a slow, cosmetically attractive repair process. Wounds involving any significant loss of dermis tend to contract and distort that part of the skin to produce scarring, and are usually treated medically by grafting.

In most parts of the body the skin has a number of appendages. These include hair of various types, sebaceous glands and sweat glands. These are important in their own right for the functions they perform; however, they are also important for another reason, and that is regeneration. These appendages

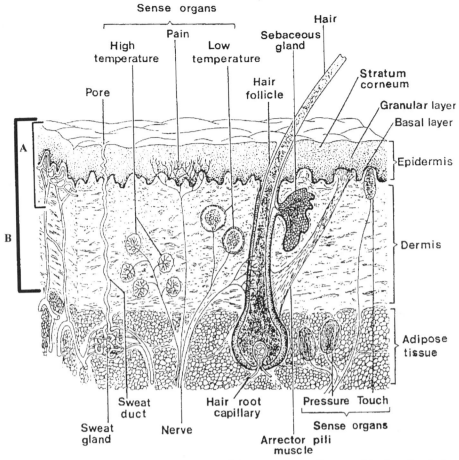

Figure 2. Diagrammatic representation of the structure of human skin showing hair follicles, sweat glands etc. as ingrowths of epidermis into the dermis
The depths A and B related to partial- or split-thickness, and full-thickness wounds, respectively. Modified from Beckett, B.S. (1982) *Biology: A Modern Introduction,* 2nd edition.

are actually ingrowths of epidermis deep into the dermis. As a consequence, a loss of dermis to a considerable depth, but which still leaves remnants of these appendages, means that regeneration of epidermis can take place promptly and, usually, successfully. The epidermis lining these ducts provides a multitude of foci from which epidermal cells can proliferate and spread out to restore the barrier function. This is perhaps best comprehended in terms of split-thickness autografting (see Figure 2). In this procedure, a layer of epidermis plus partial-thickness dermis is taken from one site on the body (e.g. thigh or buttock) and is grafted on to another site where there has been extensive and full-thickness loss of dermis and epidermis. Such a graft provides cover and restoration of barrier function and, under favourable conditions, the autograft 'takes'. The donor site is left with the remnants of the epidermal appendages, from which regeneration usually takes place successfully with no more than temporary discomfort to the patient. In severe burn cases these donor sites may have to be used repeatedly.

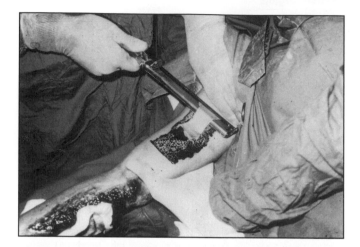

Figure 3. Split-thickness grafts with meshing
(*a*) Shows split-thickness skin being taken from a donor site.

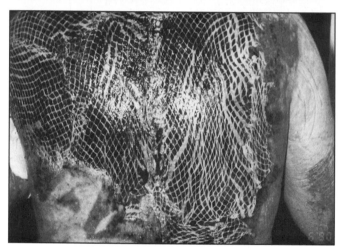

(*b*) This may then be 'meshed' to expand its area and placed on the burned patient where there has been extensive loss of skin.

Various ingenious means have been devised to 'stretch' the grafting procedure further. 'Meshing' the grafted split-thickness skin involves making a series of small vertical cuts in it so that it can indeed be stretched to cover a larger area than that from which it originated (Figure 3). The Chinese method of doing things involves chopping the split-thickness graft into tiny pieces[3]. The pieces should be smaller than 1 mm^3 and they are then immersed in saline in which they eventually float epithelial side up. After this they are spread on to a silk cloth for eventual transfer to the patient. Good results were reported in eight patients with up to 44% of the body surface covered by this method of grafting. As well as being economical with the autograft material, the procedure is said to produce less scarring.

In all of these procedures autografting is used; allografting from another individual with split-thickness skin, whether chopped up or not, is not successful because of immunological rejection[4]. The idea of giving immuno-suppressive drugs to already severely compromised burn patients is obviously not sensible, but, as we shall see, the dismantling of the epidermis and its

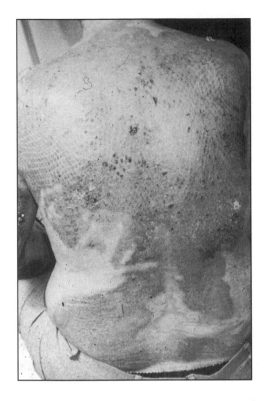

Figure 3. (contd.)
(c) The same patient several months after treatment. Photos courtesy of Dr P. Dziewlski, Burns Unit, Pinderfields Hospital, Wakefield.

reconstruction can lead to a structure that is usable. In fact, in some severe burn cases, cadaver skin is used to cover areas of skin loss. This restores the skin's barrier function for a time, although the graft is ultimately rejected after a few weeks. Nevertheless, these few weeks may be sufficient to allow the patient to stablilize and come out of shock, and generally to be in a position for autografting from donor sites to begin.

Many types of dressing have been devised for covering extensive wounds and restoring the barrier function: some purely synthetic and some constructed of biological materials. Ultimately, a living epidermis needs to be in place for a patient to be restored to anything like normal health. A possible strategy here is to provide coverings that restore the barrier function while permitting regrowth of epidermis. However, the basic fact remains that when there has been full-thickness skin loss (i.e. to below the depth of the epidermal appendages) the rate of ingrowth of epidermis from the margin of the wound is about 0.3 mm per day. It therefore takes a very long time to re-cover even a few square centimetres of full-thickness wound, even under favourable circumstances.

Another possible strategy is to graft using sheets of allogeneic keratinocytes that have lost the ability to provoke an immune response from the recipient[5]. Keratinocytes may be grown *in vitro* to form sheets, and, as will be shown in the next section, there is the possibility of them being used as allografts[6,7]. First, we consider the immunological background.

Skin immunology

A number of antigens are expressed on cells which enable the immune system to differentiate its own cells ('self') from foreign cells[8]. These are the major histocompatibility complex (MHC) antigens in humans, also called human leukocyte antigens (HLAs). MHC class I antigens are also called HLA-A, -B and -C, and the MHC II antigen is called HLA-D. MHC class I antigens are expressed on all nucleated cells, whereas the MHC class II antigens are limited to a few cell types, such as Langerhans cells and basal keratinocytes.

Langerhans cells, which are resident in the epidermis, are part of the specific immune system. They are able to detect and process antigens entering the skin and they then migrate to the secondary lymphoid tissues where they present the antigens with their autologous MHC class II molecules to lymphocytes. Langerhans cells also provide the second signal for the activation of helper T-cells which can then activate specific B-lymphocytes by secreting cytokines, although Langerhans cells seem not to be required for activation of cytotoxic T-lymphocytes[9].

It is thought that when allogeneic skin is placed onto a wound bed, blood vessels from the recipient grow into the graft and join (inosculate) with the cut ends of the allogeneic vasculature of the graft. Langerhans cells in the graft come into contact with material from the recipient, their MHC class II molecules are perceived as foreign, and an immune response to the graft is therefore initiated. Despite complete MHC matching and negative mixed leucocyte reactions, whole skin allografts are rejected in humans.

In contrast to normal basal keratinocytes *in vivo*, cultured human keratinocytes do not express MHC class II antigens[10] and fail to stimulate autologous lymphocytes in the mixed lymphocyte reaction[11]. Furthermore, Langerhans cells do not survive in culture conditions suitable for keratinocytes, partly due to the presence of glucocorticosteroids in the culture medium. It has been proposed that the loss of Langerhans cells makes it possible for allogeneic keratinocyte sheets to be used as grafts without eliciting an acute immune response[7].

Growing keratinocyte sheets

Culture of epidermal keratinocytes *in vitro* is now possible but rather difficult. Early studies were carried out on outgrowths of epidermis from explanted biopsies but, in 1975, Rheinwald and Green published a method for growing isolated keratinocytes in culture. There are only a limited number of cells capable of proliferating in intact epidermis and it is important to maximize the initial harvesting of these cells as well as their subsequent growth rates. The Rheinwald and Green technique[12], which is still the one most commonly used today, involves co-culture of keratinocytes isolated from a sample of epidermis by enzyme treatment, with lethally irradiated 3T3 fibroblasts, which supply the various growth and attachment factors needed by keratinocytes. Eventually the keratinocytes form a monolayer and may begin to stratify, and the fibroblasts die off (Figure 4). Before the keratinocytes reach confluence, the culture may be split and expanded to produce sheets of epidermis of a considerable area. Expansions of 10 000-fold are routinely achievable. Such sheets of epidermis have been used quite extensively as epidermal grafts in severely burned patients[13,14] and for treating leg ulcers (Figure 5)[15,16].

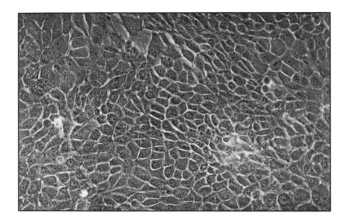

Figure 4. Sheet of human keratinocytes cultured in the laboratory

Keratinocytes may have two modes of growth[12,17]. One is to proliferate, and this is the mode of behaviour of some of the cells of the basal layer of the epidermis. New daughter cells are constantly being produced, and they then move outwards. As they do so, they enter the second mode, i.e. they begin a process of terminal differentiation. Keratinocytes produce large amounts of the intermediate filament protein keratin and the types of keratin polypeptide produced depend on the stage of differentiation[18,19]. Keratins K5 and K14 are expressed in the basal layer but, as the cells differentiate, K1 and K10 are produced, and soon most of the cell organelles start to disappear. Transglutaminase is produced or activated near the cell membrane and a number of proteins become cross-linked at the cell membrane, resulting in the formation of an inert 'cornified envelope' around the cell or corneocyte, which by this time is little more than a tough, flattened bag of keratin intermediate filaments. These dead cells of the cornified layer stick together as 'squames' and represent the first line of the epidermal barrier. They are gradually sloughed off by abrasion and are replaced from below. The transit time from basal cell dividing to a corneocyte sloughing off as a squame is about 28 days for normal human epidermis.

Keratinocytes are rather fastidious in their nutritional requirements[20]. The growth medium needs to contain additives, such as cholera toxin, insulin, hydrocortisone, epidermal growth factor (EGF), transferrin and tri-iodothyronine, as well as fetal calf serum, to achieve a proliferative population of cells. The signals that cause a change from proliferative mode to differentiating mode are poorly understood, although a change in the Ca^{2+} concentration in the medium can cause the cells to make this switch[20]. This balance is quite delicate and once the keratinocytes have entered the differentiating mode the culture will stop proliferating. Careful control of the conditions, and precise choice of the time of splitting the culture, are very important in the production of sheets of keratinocytes for clinical use as biological dressings. Nevertheless, areas of 1 m^2 or more of keratinocyte sheet may be produced from a very small biopsy over a period of several weeks.

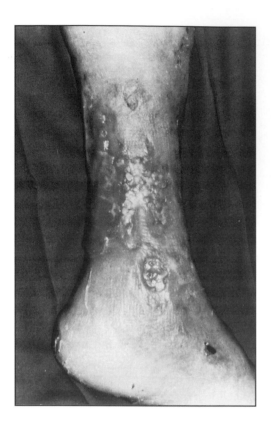

Figure 5. Typical appearance of a leg ulcer

Recently, some so-called defined keratinocyte growth media have appeared[21,22]. They are 'defined' in the sense of not requiring the addition of fetal calf serum; however, it is necessary to add a saline extract of bovine pituitary instead, so 'defined' has a rather limited meaning.

It may be noted here that cultured keratinocyte sheets do not usually contain the other epidermal cell types, such as Langerhans cells and melanocytes, because the culture conditions are chosen to support proliferative keratinocyte growth specifically.

Is allografting with keratinocyte sheets successful?

Cultured keratinocyte sheets (autografts) were used in 1984 to graft two children who had lost over 95% of their skin in a fire[6] and this surgical procedure was successful, although not cosmetically very attractive. Subsequently, cultured allograft keratinocyte sheets have been used to treat over 130 burn patients in Europe, Japan, Mexico and the U.S.A.[14,16] Keratinocyte sheets have also been used to treat graft donor sites, tattoo excisions, leg ulcers that fail to heal[23] and giant congenital naevi excision.

There are two aspects to this that warrant closer examination. These are, firstly, to ask whether the procedure is successful in stimulating wound healing (which it is), and, secondly, to ask whether the grafted allogeneic keratinocytes persist. This second question is somewhat controversial. The evidence has come from grafting with sex-mismatched donor keratinocytes, and later testing for allogenic keratinocytes using a probe specific for the Y-chromosome[24]. DNA fingerprinting, mismatched A and B blood-group antigens and MHC class I antigens have also been used to investigate the survival of allogeneic cells[25,26]. There is controversy as to how long allogeneic cells persist; using DNA fingerprinting and a probe for the Y-chromosome, the allogeneic cells could not be detected after 1 week.

It was concluded that the allografts do not 'take' and that the graft merely acts as a biological dressing and is rapidly replaced by autologous keratinocytes. However, Yang-bing and co-workers[25] have detected allogeneic grafted cells 10, 35 and 92 days post-grafting using antibodies to stain for the A or B blood antigen or detecting the Y-chromosome by polymerase chain reaction (PCR). This discrepancy might be explained by the type of wound bed to which the allografts were applied. Yang-bing and co-workers placed allografts onto split-thickness graft donor sites where there was apparent clinical take, whereas van der Merwe and co-workers[26] placed allografts onto deep burned wounds. The take of epidermal allografts is known to be much less successful on to connective tissue, fat tissue and granulation tissue than on to dermis itself[14]. Allogeneic cells are believed to be replaced progressively by autologous cells from epidermal appendages[25].

Cultured keratinocyte allografts survived as long as matched allogeneic split-thickness skin grafts, but all were rejected after 10 to 22 days[4]. At the time of rejection there was a 10-fold increase in circulating donor-specific cytotoxic T-cells that resulted from clonal expansion in response to antigen. Mauduit and co-workers[27] also studied in humans the immune response to cultured keratinocyte allografts applied to leg ulcers. The presence of circulating anti-class I MHC antibodies in the grafted patients was not detected. These differences may also have been due to the degree of take, which may depend on the nature of the wound bed. In many species of animal, allogeneic skin grafts are rejected much more fiercely than other organ grafts over the same histocompatibility barrier[28]. The reason why rejection with skin grafts occurs with such vigour may be due to the keratinocytes themselves which secrete numerous cytokines, such as interleukin (IL)-1, IL-3 and IL-6, which may greatly augment the immune response[29,30].

Skin banking and the HIV problem

The technique of grafting with sheets of allogeneic keratinocytes seems to be sufficiently successful to warrant consideration of the idea of a 'skin bank' with cryopreserved sheets of epidermis[31]. There is a problem at the present time that any keratinocytes used in such a procedure should be screened to be free from human immunodeficiency virus (HIV), as there have been some reports that HIV can be transferred in skin. Possibly this is via Langerhans cells[32], but there is sufficient uncertainty that health authorities insist on screening the material. At present, in the U.K., permission will not be given for the grafting procedure unless the donor of the keratinocytes has been screened at the time of donation and again after 90 days (in case seroconversion had not taken place at the time of donation). This makes cryopreservation an even more desirable aim, because cells must be grown and then held for 90 days to allow for the second test. This is the present situation, although it might change if a PCR test for HIV nucleic acid actually in the epidermal sample were to become accepted by health authorities.

Cryopreservation

Cryopreservation of sheets of epidermis is possible provided that the cell sheets are first impregnated with either glycerol or dimethylsulphoxide (DMSO) as cryoprotectants[33,34]. Usually the permeation with the cryoprotectant (5–10%) is carried out in the presence of fetal calf serum, at 4 °C. Subsequently, the material is cooled slowly at, say, –1 °C per minute to –70 °C, and then transferred to liquid nitrogen for storage (Figure 6). DMSO seems to be preferable and certainly permeates more rapidly than glycerol.

The viability of cells in such sheets may be estimated on re-warming after several months of storage by measuring, for example, lactate dehydrogenase leakage. With careful attention to detail, 35–55% recovery of viable cells can normally be achieved.

Dermal equivalents and skin equivalents

Although grafting with sheets of keratinocytes is feasible (and even if the cells do not actually persist on the wound they seem to stimulate wound healing), a sheet of keratinocytes is a very thin and mechanically weak structure. As well as stimulating the wound healing process, it does of course restore the skin's barrier function to some extent. However, the result, especially for deep wounds with extensive loss of tissue, is cosmetically poor and is likely to get worse with time as the wound undergoes contraction. There are, therefore, good reasons for trying to replace dermis as well as epidermis, in other words aiming at a more complete reconstruction of skin from its component parts.

Fibroblasts

The chief cell type of the dermis is the dermal fibroblast, and such cells are readily grown in culture as monolayers. The fibroblasts have numerous roles in the dermis. These include producing collagen, elastin, glycosaminoglycans

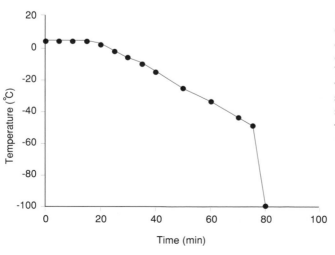

Figure 6.
Cryopreservation of epidermis or sheets of keratinocytes
The sheets are treated with a cryopreservative and then subjected to slow cooling until about –40 °C.

and other extracellular matrix proteins. However, they also produce, or are capable of producing, enzymes such as collagenase and other proteases that 're-model' deposited collagen, a process that is very important in the later stages of the wound-healing process. In fact, collagenases are members of the group of tissue metallo-proteinases whose control and activation are very complex and not well-understood but are being intensively studied[35].

The contrast between dermis and epidermis should be recalled here. Whereas epidermis is practically all cells, the dermis is a much more complicated structure. It consists of a matrix of predominantly collagen, with some elastin and glycosaminoglycans, with fibroblasts distributed through it, adhering to the collagen fibres. In addition to the ingrowths of epidermis (hair follicles, sweat glands) mentioned earlier, dermis contains blood vessels, nerves and so on (see Figure 2). It is important to realize that fibroblasts in dermis are growing in a meshwork of collagen fibres rather than as a monolayer, which is how they are usually grown in culture. Such a collagen-based structure may, to some extent, be reconstructed in the laboratory[36], and it is noticeable that a number of facets of fibroblast behaviour, including their responses to growth factors, are significantly different in this situation compared with that in monolayers[37]. Culturing fibroblasts in collagen gels or 'lattices' produces structures known as dermal equivalents[38] (see below) that have a potential use for filling extensive wounds and form a basis for constructing skin equivalents (Figure 7).

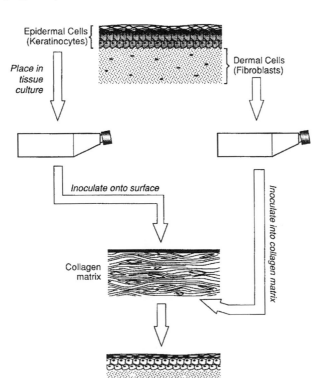

Epidermal Cells (Keratinocytes)

Place in tissue culture

Dermal Cells (Fibroblasts)

Inoculate onto surface

Inoculate into collagen matrix

Collagen matrix

Figure 7. Overall scheme for reconstitution of skin from skin cells (fibroblasts and keratinocytes) and a collagen lattice

Dermal equivalents

A dermal equivalent is constructed by seeding dermal fibroblasts into a re-formed, 3-dimensional collagen matrix, and a skin equivalent is produced by layering keratinocytes on the upper surface of such a structure. It will be noted immediately that such structures are highly simplified when compared with skin itself; nevertheless, they represent the first steps in skin reconstruction, and the model is clearly capable of development to produce more complex structures.

The idea of a reconstructed skin along these lines was first proposed by Bell and co-workers[36] and patents for organ reconstruction were taken out. Since that time a number of important developments have taken place. Not only are such structures of great interest for their clinical potential in the wound-healing situation[39,40], but, it was soon realized, they also have potential for studies on skin-cell biochemistry, as well as the possibility of being used as substitutes for animals in testing drugs and cosmetics for efficacy or toxicity. At least one commercial company offered skin equivalent systems for such testing. Following is a description of how dermal and skin equivalents may be constructed and of their potential for further levels of complexity to be added. It is not too difficult to see that such structures might also be constructed with other cell types and extracellular matrix (ECM) components such that organs other than skin might be fabricated[41].

The most common basis for constructing a dermal equivalent is acetic-acid-solubilized, native, rat tail type-I collagen, although other collagen such as calf skin collagen may also be used. This material is polymerized by bringing the pH to neutrality and simultaneously adding dermal fibroblasts and the appropriate growth medium. The collagen forms a gel, in sterile conditions, in about 30 min, and thereafter the gel will contract in thickness and diameter over a period of days (Figure 8). This process may, superficially at least, represent a model for studying the wound contraction process. The collagen gel may contract to, say, 20% of its original diameter and at this stage takes on

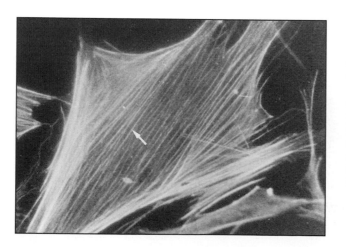

Figure 8. Fibroblasts in a dermal equivalent generate a tension
Cells were stained with rhodamine–phalloidin and photographed in the fluorescence microscope. The actin fibres are clearly visible (arrow). Photograph kindly supplied by P. Stephens.

properties that make it more like dermis than collagen gel, although the collagen concentration is lower than in dermis (Figure 9). It becomes tough, easy to handle, and opaque — resembling dermis — although it should be remembered that, at least initially, it lacks elastin and glycosaminoglycans, let alone any of the other cells and structures found in dermis.

The process of contraction of dermal equivalents has been extensively studied[36,38]. The kinetics of the process are quite well-defined and, more recently, it has been observed that fibroblasts may form a continuous circle at the periphery of the (usually circular) structure. Staining these fibroblasts with rhodamine–phalloidin reveals a continuous 'cable' of actin filaments that is presumably responsible for generating the tension that causes the contraction (Figure 10). This is somewhat reminiscent of the 'purse-string' structure that has been observed at the periphery of healing embryonic wounds[42]. Dermal equivalents clearly have the potential for incorporating within them other ECM components, such as hyaluronan, glycosaminoglycans and so on, with the aim of making them relatively closer in structure to true dermis. They represent structures that could potentially be used to fill a wound space in an extensive wound where there has been considerable loss of tissue. As has been noted, dermis can regenerate, but wound contraction typically takes place with unsightly effects. Filling the wound with a dermal equivalent type of structure may, to some extent, prevent this. The next process that needs to occur is angiogenesis, the growth and infiltration into the dermal equivalent of new blood vessels. Such a process has so far been relatively little studied. It is vital because of the question of oxygen and nutrient supply to the 'tissue'[43] and, indeed, it would mimic the process of granulation tissue formation in normal wound healing. Also, of course, it would represent an attractive model system for the study *in vitro* of angiogenesis and angiogenic factors[44].

Figure 9. Contraction of a dermal equivalent over 21 days in culture
The contracted dermal equivalent has a tissue-like structure.

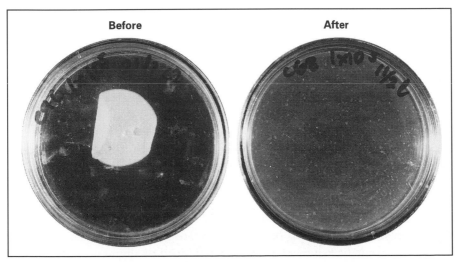

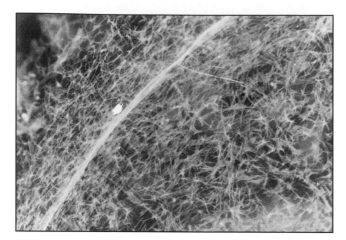

Figure 10. At the peiphery of a dermal equivalent undergoing contraction, fibro-blasts form a 'purse-string' structure
This dermal equivalent was stained with a rhodamine phalloidin (see Figure 8). Photograph kindly supplied by P.E. Genever.

Skin equivalents

A dermal equivalent may be transformed into a skin equivalent by placing keratinocytes on its upper surface. This process often causes further contraction and some care needs to be exercised to obtain a sufficient area of reconstructed skin for grafting.

An important aspect of this reconstruction is to achieve something approaching a normal epidermis on top of the dermal equivalent. The keratinocytes adhere and grow to form a confluent monolayer initially. However, it has been found that at this stage it is advantageous to raise the whole structure out of the culture medium so that the epidermis is air-exposed, while the dermis is submerged in the nutrient medium (Figure 11). The epidermis is then in its 'natural habitat' and is fed from below via the dermis. Under these circumstances a good histology will develop, with differentiation of the keratinocytes (Figure 12), and this material is potentially useful for grafting or for testing dermatological products.

At present this technique is at the stage of being viable but expensive. For covering a small area, such as a leg ulcer or a tattoo removal, a suitable area of skin equivalent may be produced quite rapidly. However, to provide cover for an individual with 50% or more surface area burns, 1 m^2 or more of material must be produced. Not only is this a considerable and somewhat time-consuming proposition, but there is also a very high cost in skilled manpower and culture materials and media.

Modified skin equivalents

Given that the technology exists to take the repair of major deep wounds to this stage, it has to be borne in mind that the skin equivalent whose construction is described above lacks pigmentation (i.e. melanocytes), as well as sweat glands, hair follicles, nerve cells, and all the other skin-associated structures. From the patients' point of view the covering may be life-saving and viable, but the cosmetic result will be poor and there would be the hazard of u.v. damage as would be experienced by a light-skinned or albino person. Thus the art of total reconstruction of skin still has some way to go.

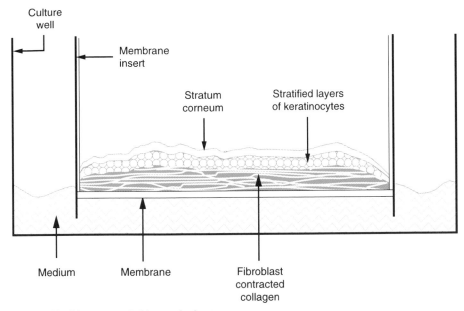

Figure 11. Air-exposed skin equivalent
A skin equivalent is created (see Figure 7) and then the artificial tissue is raised up so that the 'dermis' is in contact with the growth medium but the 'epidermis' is exposed to air.

Steps are already being taken to incorporate melanocytes into skin equivalents. Boyce and colleagues[45] reported some progress in this direction. They incorporated human melanocytes into a skin equivalent system and then grafted them on to excised skin wounds on athymic mice (Figure 13). Several medium compositions were tried in producing the composite skin equivalents because optimal media compositions for growing keratinocytes and melanocytes are rather different from one another. It was observed that the incorporation of melanocytes caused a statistically significant reduction in the

Figure 12. Histology of normal skin (*left*) and a skin equivalent (*right*) side-by-side for comparison
The upper layer of dead corneocytes (stratum corneum) is clearly visible in each case, indicating that in the skin equivalent the keratinocytes undergo terminal differentiation. Photograph courtesy of J.M. Page.

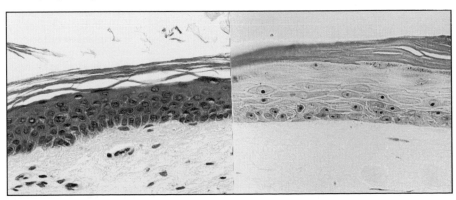

degree of wound contraction. The grafts on to mice survived for at least 6 weeks and showed full pigmentation. Furthermore, electron microscopy showed the presence of melanocytes, melanosomes and pigment transfer to keratinocytes in the pigmented skin. These experiments are important because they set the scene for overcoming the hypopigmentation problem. This would be especially important when grafting skin equivalents on to dark-skinned individuals.

The problem of hair follicles, sweat glands and other epidermal ingrowths seems to be more difficult to solve. These structures are formed at the embryonic stage and their number in the skin does not appear to change throughout life. With present technology, therefore, there would seem to be no chance of their being formed *de novo* by ingrowths of epidermis into the dermis.

The future

Knowledge of skin cell biology, cell–matrix interactions, and the influence of cytokines and other growth factors is increasing at a rapid pace and will be applied to extend the possibilities of skin reconstruction, essentially a branch of what has been called 'tissue engineering'[46,47]. As knowledge accumulates, one will expect similar strategies and techniques to be applied to other tissues such as bone, cartilage and blood vessels[41,48].

Gene therapy

Another aspect of skin cell technology, however, is in the area of gene therapy. The cultivatability of skin cells (keratinocytes or fibroblasts) in principle makes them a suitable target for genetic manipulation. Furthermore, the accessibility of the skin for obtaining cell samples in the first place, and re-implanting them after gene therapy by minor surgical procedures, makes the skin an attractive target. Techniques for transfecting cells are developing rapidly and, indeed, becoming more efficient. The sorts of therapy that might be imagined initially would be to provide a soluble enzyme or factor that an

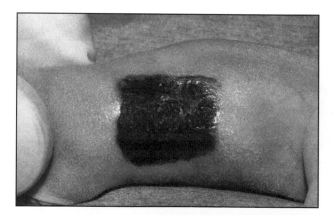

Figure 13. Cultured skin equivalent containing human melanocytes transplanted onto athymic mouse shown 6 months after grafting Photograph kindly supplied by S.T. Boyce and reproduced from reference 45 with permission.

individual was lacking, through the skin route. Obviously, a factor such as a clotting factor that was needed in constant low levels would be the easiest to provide. Other factors that vary with physiological state, such as insulin, would be more difficult because control mechanisms would have to be built in.

The skin provides a very attractive route for gene therapy. Either keratinocytes or fibroblasts may be transformed with a gene and the cells replaced to provide the gene product in a genetically deficient individual. The amount of surgery required for such procedures is minimal, but there are a number of technical problems associated with this technology. If the gene product supplied to the deficient donor via the transplanted cells is a secreted one then it may trigger an immune response, and the cells may be destroyed. Morgan and colleagues had reported in 1987, for example, the expression of exogenous growth hormone gene by cultured keratinocytes[49]. These keratinocytes were grafted as an epithelial sheet on to athymic mice and continued to produce growth hormone for a few days.

Another problem is that, since epidermis is avascular, any secreted protein must traverse the basement membrane that separates epidermis from dermis in order to reach the circulation. Morgan and colleagues were able to show the secretion of biologically active human growth hormone by the keratinocytes, but were unable to determine the rate of diffusion of the hormone from the graft site to the bloodstream because of the low sensitivity of the radio-immunoassay used[49].

In 1989 Fenjves and colleagues reported the systemic distribution of human apolipoprotein E (apoE) secreted by grafts of sheets of human epidermal keratinocytes placed on athymic mice[50]. (It was known that cultured keratinocytes synthesize and secrete apoE.) This indicated that a protein as large as apoE (299 amino acids) can traverse the epidermal–dermal barrier and achieve systemic distribution. The keratinocyte layer was placed on prepared sites on the recipient animals. Immunological problems in these studies were obviated by using athymic mice, but problems of rejection had been encountered in previously reported experiments. Other workers have used transfected fibroblasts as the vehicle for delivery of the deficient gene but, again, immunological problems caused the gene-product expression to cease after about a month.

In an attempt to define more precisely some of the problems, Palmer and co-workers (1991) transplanted into rats fibroblasts transformed using retroviral vectors to contain the gene for adenosine deaminase (ADA)[51]. This is an intracellular protein and, therefore, should not be immunogenic, in contrast with a secreted protein such as factor IX that one might wish to use in therapy. They also stressed the importance of using primary fibroblasts rather than immortalized ones (which had been used fairly successfully in a number of previous studies), which grow uncontrollably and may form tumours in recipients.

Palmer and colleagues seeded the retrovirally transformed fibroblasts into collagen matrices (i.e. dermal equivalents) which were then placed in circular, full-thickness skin wounds in rats and protected with a collagen/chondroitin sulphate matrix dressing. Initial enzyme activities in the transformed fibroblasts were so high that $\sim 5 \times 10^7$ cells would have provided sufficient ADA activity to correct the defect in a human with ADA deficiency. This number of cells is a feasible number to transplant.

It was found that ADA activity persisted for about 1 month but then fell off very dramatically. However, this was apparently not an immunological effect, because the transplanted cells survived for at least 8 months, but rather was sequence-specific gene inactivation.

From this limited number of experiments it seems as though it will soon be feasible to transplant, or rather re-implant, retrovirally transformed skin cells on to human patients with genetic deficiencies of certain gene products[52], although the immunological problems and those connected with long-term expression of the vector-encoded product must be solved. Next to injection of transformed cells, a small transplant of skin cells, epidermal keratinocyte sheets or fibroblasts in a collagen gel, would seem to be the least traumatic route.

Conclusions

Knowledge of skin cell biology has allowed the reconstruction of human skin for grafting and other purposes and one can expect further developments in the technology to be rapid. The number of potential recipients for such skin grafts is very large. Langer and Vacanti[47] reported that each year in the U.S.A. some 2150000 individuals undergo surgical procedures for burns: 150000 of these are hospitalized and 10000 die[47]. In addition, in 1993 there were 1500000 treatments for pressure sores, 500000 for venous stasis ulcers and 600000 for diabetic ulcers, and the numbers in the U.K. and Europe must be similar *pro rata*. As with all 'advanced' medicine, there are always more patients than the cost structure of the hospital system can allow and, inevitably, the relative costs of such novel treatments must be weighed against the cost-benefits of conventional treatments, such as normal autografting or the use of synthetic materials in dressings and implants. Finally, the potential for gene therapy will surely give fresh impetus to understanding the cell biology of transplantation rejection and sustained delivery of a gene product.

References

1. Clark, R.A.F. (1989) Wound repair. *Curr. Opin. Cell Biol.* 1, 1000–1008
2. Odland, G.F. (1983) Structure of the skin, in *Biochemistry and Physiology of the Skin* (Goldsmith, L.A., ed.), pp. 3–63, Oxford University Press, New York
3. Zhang, M.-L., Wang, C.-Y., Chang, Z.-D., Cao, D.-X. & Han, X. (1986) Microskin grafting: II clinical report. *Burns* 12, 544–548

4.   Auböck, J. & Fritsch, P. (1987) Epidermal allografts in humans: an unattainable dream? *Dermatologica* **175**, 161–165

5.   Faure, M., Mauduit, G. Schmitt, D., Kanitakis, J., Demidem, A. & Thivolet, J. (1987) Growth and differentiation of human epidermal cultures used as auto- and allografts in humans. *Br. J. Dermatol.* **116**, 161–170

6.   Green, H., Kehinde, O. & Thomas J. (1979) Growth of cultured human epidermal cells into multiple epithelia suitable for grafting. *Proc. Natl. Acad. Sci. U.S.A.* **76**, 5665–5668

7.   Gallico, G.G., O'Connor, N.E., Compton, C.C., Kehinde, O. & Green, H. (1984) Permanent coverage of large burn wounds with autologous human cultured epithelium. *N. Engl. J. Med.* **311**, 448–451

8.   Jonker, M., Hoogeboom, J., Van Leuuwen, A., Koch, C.T., Van Oud Alblas, D.B. & Van Rood, J.J. (1979) Influence of matching for HLA-DR antigens on skin graft survival. *Transplantation* **27**, 91–94

9.   Peeler, J.S. & Niederkorn, J.K. (1986) Antigen presentation by Langerhans cells *in vivo*: donor-derived Ia+Langerhans cells are required for induction of delayed-type hypersensitivity but not for cytotoxic T-lymphocyte responses of alloantigens. *J. Immunol.* **136**, 4362–4371

10. Morhenn, V.B., Benike, C.V., Cox, A.J., Charron, D.J. & Engelman, E.G. (1982) Cultured human epidermal cells do not synthesise HLA-DR. *J. Invest. Dermatol.* **78**, 32–37

11. Hefton, J.M., Amberson, J.B., Biozes, D.G. & Weksler, M.E. (1981) Human epidermal cells no longer stimulate allogeneic lymphocytes after growth in culture. *J. Invest. Dermatol.* **76**, 308–309

12. Rheinwald, J.G. and Green, H. (1975) Serial cultivation of strains of human epidermal keratinocytes: the formation of keratinizing colonies from single cells. *Cell* **6**, 331–344

13. O'Connor, N.E., Mulliken, J.B., Banks-Schlegal, S., Kehinde, O. & Green, H. (1981) Grafting of burns with cultured epithelium prepared from autologous epidermal cells. *Lancet* **i**, 75–78

14. Madden, M.R., Finkelstein, J.L., Staiano-Coico, L., Goodwin, C.W., Shires, T., Nolan, E.E. & Hefton, J.M. (1986) Grafting of cultured allogeneic epidermis on second- and third-degree burn wounds on 26 patients. *J. Trauma* **26**, 955–960

15. Harris, I.R., Bottomley, W., Wood, E.J. & Cunliffe, W.J. (1993) Use of autografts for the treatment of leg ulcers in elderly patients. *Clin. Exp. Dermatol.* **18**, 417–420

16. Phillips, T.J. & Gilchrest, B.A. (1992) Clinical applications of cultured epithelium. *Epithelial Cell Biol.* **1**, 39–46

17. Potten, C.S., Wichmann, H.E., Loeffler, M., Dobek, K. & Major, D. (1982) Evidence for discrete cell kinetic subpopulation in mouse epidermis based on mathematical analysis. *Cell Tissue Kinet.* **15**, 305–320

18. Steinert, P.M., Steven, A.C. & Roop, D.R. (1985) The molecular biology of intermediate filaments. *Cell* **42**, 411–419

19. Sun, T.-T., Eichner, R., Nelson, W.G., Tseng, S.C., Weiss, R.A., Jarvinen, M. & Woodcock-Mitchell, J. (1983) Keratin classes: molecular markers for different types of epithelial differentiation. *J. Invest. Dermatol.* **81**, 109s–115s

20. Watt, F.M. (1988) The epidermal keratinocyte. *BioEssays* **8**, 163–167

21. Wilke, M.S., Edams, M. & Scott, R.E. (1988) Ability of normal human keratinocytes that grow in culture in serum-free medium to be derived from basal cells. *J. Natl. Cancer Inst.* **80**, 1299–1304

22. Johnson, E.W., Meunier, S.F., Roy, C.J. & Parenteau, N.L. (1992) A new method of keratinocyte cultivation: maintenance of the proliferative cell population in a defined environment. *J. Invest. Dermatol.* **96**, 561

23. Leigh, I.M., Purkis, P.E., Navsaria, H.A. & Phillips, T.J. (1987) Treatment of chronic venous leg ulcers with sheets of cultured allogeneic keratinocytes. *Br. J. Dermatol.* **117**, 591–597

24. Burt, A.M., Pallett, C.D., Sloane, J.P., O'Hare, M.J., Schafler, K.F., Yardeni, P., Eldad, A., Clarke, J.A. & Gusterson, B.A. (1989) Survival of cultured allografts in patients with burns assessed with probe specific for Y chromosome. *Br. Med. J.* **298**, 915–917

25. Yang-Bing, Z., Xiong-Fei, Z., Ao(Ngao), L., Shu-Zhen, L., Xu, W., Shu-Zhen, H. & Xia-Ti, Z. (1992) Clinical observations and methods for identifying the existence of cultured epidermal allografts. *Burns* **18**, 4–8

26. Van der Merwe, A.E., Mattheyse, F.J., Bedford, M., Van Helden, P.D. & Rossouw, D.J. (1990) Allografted keratinocytes used to accelerate the treatment of burn wounds are replaced by recipient cells. *Burns* **16**, 193–197

27. Mauduit, G., Faure, M., Demidem, A., Kanitakis, J. & Thivolet, J. (1986) Cultured human epidermis used as allografts: studies on their differentiation *in vivo. J. Invest. Dermatol.* **87**, 154

28. Fabre, J.W. & Morris, P.J. (1975) Studies on the specific suppression of renal allograft rejection in presensitised rats. *Transplantation* **19**, 121–133

29. Fabre, J.W. (1988) 'Immune' functions of parenchymal cells might contribute to their susceptibility to rejection. *Transplant Int.* **1**, 165–167

30. Barker, J.N.W.N., Mitra, R.S., Griffiths, C.E.M., Dixit, V.M. & Nickoloff, B.J. (1991) Keratinocytes as initiators of inflammation. *Lancet* **337**, 211–214

31. Kearney, J.N. (1991) Cryopreservation of cultured skin cells. *Burns* **17**, 380–383

32. Cameron, P.U., Freudenthal, P.S., Barker, J.D., Gezelter, S., Inaba, K. & Steinman, R.M. (1992) Dendritic cells exposed to human immunodeficiency virus type 1 transmit a vigorous cytopathic infection to CD4[+] T cells. *Science* **257**, 383–387

33. Roseeuw, D., De Coninck, A., Neven, A.M., Vandenberghe, Y., Kets, E. Verleye, G. & Rogiers, V. (1991) Fresh and cryopreserved cultured keratinocyte allografts for wound healing. *Toxicol. in Vitro* **5**, 579–583

34. Teepe, R.G.C., Koebrugge, E.J., Ponec, M. & Vermeer, B.J. (1990) Fresh versus cryopreserved cultured allografts for the treatment of chronic skin ulcers. *Br. J. Dermatol.* **122**, 81–89

35. Matrisian, L.M. (1990) Metalloproteinases and their inhibitors in matrix remodelling. *Trends Genet.* **6**, 121–125

36. Bell, E., Ivarrson, B. & Merrill, C. (1979) Production of a tissue-like structure by contraction of collagen lattices by human fibroblasts of different proliferative potential *in vitro. Proc. Natl. Acad. Sci. U.S.A.* **76**, 1247–1278

37. Jutley, J.K., Wood, E.J. & Cunliffe, W.J. (1993) Influence of retinoic acid and TGF-β on dermal fibroblast proliferation and collagen production in monolayer cultures and dermal equivalents. *Matrix* **13**, 235–241

38. Rowling, P.J.E., Raxworthy, M.J., Wood, E.J., Kearney, J.N. and Cunliffe, W.J. (1990) Fabrication and reorganisation of dermal equivalents suitable for skin grafting after major cutaneous injury. *Biomaterials* **11**, 181–185.

39. Grinnell, F. (1994) Fibroblasts, myoblasts and wound contraction. *J. Cell Biol.* **124**, 401–404

40. Wassermann, D., Schlotterer, M., Toulon, A., Cazalet, C., Marien, M., Cherruau, B. & Jaffray, P. (1988) Preliminary clinical studies of a biological skin equivalent in burned patients. *Burns* **14**, 326–330

41. Haynes, S.L., Kearney, J.N., Davies, G.A., Wood, E.J. & Fisher, J. (1991) Interactions of vascular smooth muscle cell with collagen matrices. *Clin. Mater.* **7**, 247–252

42. Martin, P. & Lewis, J. (1991) Actin cables and epidermal movement in embryonic wound healing. *Nature (London)* **360**, 179–183

43. Silver, I.A. (1969) The measurement of oxygen tension in healing tissue. *Prog. Resp. Res.* **3**, 124–135

44. Schreiber, A.B., Winkler, M.E. & Derynck, R. (1986) Transforming growth factor-α; a more potent angiogenic mediator than epidermal growth factor. *Science* **232**, 1250–1253

45. Boyce, S.T., Medrano, E.E., Abdel-Malek, Z., Supp, A.P., Dodick, J.M., Nordlund, J.J. & Warden, G.D. (1993) Pigmentation and inhibition of wound contraction by cultured skin substitutes with adult melanocytes after transplantation to athymic mice. *J. Invest. Dermatol.* **100**, 360–365

46. Bell, E. (1991) Tissue engineering: a perspective. *J. Cell Biochem.* **45**, 239–241

47. Langer, R. & Vacanti, J.P. (1993) Tissue engineering. *Science* **260**, 920–926

48. Peppas, N.A. & Langer, R. (1994) New challenges in biomaterials. *Science* **263**, 1715–1720

49. Morgan, J.R., Barrandon, Y., Green, H. & Mulligan, R.C. (1987) Expression of an exogenous growth hormone gene by transplantable human epidermal cells. *Science* **237**, 1476–1479

50. Fenjves, E.S., Gordon, D.A., Pershing, L.K., Williams, D.L. & Taichman, L.B. (1989) Systemic distribution of apolipoprotein E secreted by grafts of epidermal keratinocytes: implications for epidermal function and gene therapy. *Proc. Natl. Acad. Sci. U.S.A.* **86**, 8803–8807

51. Palmer, T.D., Rosman, G.J., Osborne, W.R.A. & Miller, A.D. (1991) Genetically modified skin fibroblasts persist long after transplantation but gradually inactivate introduced genes. *Proc. Natl. Acad. Sci. U.S.A.* **88**, 1330–1334

52. Mulligan, R.C. (1993) The basic science of gene therapy. *Science* **260**, 926–932

5

Opsin genes

B. Edward H. Maden

Department of Biochemistry, University of Liverpool, PO Box 147, Liverpoool L69 3BX, U.K.

Introduction

The retina is the interface between the external world and our visual perception of it. Light, focused by the lens onto the retina, excites in a quantized manner photoreceptor molecules in specialized cells called rods and cones. In rods the photoreceptor is rhodopsin.

Rhodopsin consists of an apoprotein, opsin, with an internally bound chromophore, 11-*cis* retinal. Rhodopsin is very abundant in rods, some 5×10^7 molecules per rod being packaged into numerous, closely stacked membranous discs. Photoexcitation of just one rhodopsin molecule is sufficient to stimulate a rod cell. This exquisite sensitivity is the basis of nocturnal vision: stimulation of just a few rods evokes perception.

Rhodopsin has been isolated from many species. In mammals its absorption maximum is close to 500 nm. The absorption spectrum of human rhodopsin in rods closely matches the dark-adapted spectral sensitivity of human subjects[1]. Because rhodopsin is the only photoreceptor for rod vision, rod vision is monochromatic and does not distinguish between colours. Colour vision is mediated by cones. Humans with normal colour vision can match any given colour by mixing appropriate intensities of three primary colours from the red, green and blue regions of the spectrum. It was long ago proposed that human colour vision depends on three classes of receptor with spectral maxima in the red, green and blue regions, respectively. This has been confirmed by various groups of workers by microspectrophotometry of single cones[2]. The absorption maxima in colour-normal subjects are as follows: blue, approx. 420 nm; green, approx. 530 nm; and red, approx. 560 nm. (See section entitled Colour vision: the tuning problem.) Red and green cones are concentrated in the fovea, where there are no rods or blue cones; all three cone types occur peripherally, although greatly outnumbered by rods.

Cones contain less of their respective pigments than do rods, and are less sensitive than rods to low levels of illumination. The transition from cone-dominated colour vision to rod-dominated monochromatic vision occurs at dusk and reverses at dawn. This is subjectively apparent to an observer far from city lights or other local illumination.

Because mammalian cone pigments are present in small quantities, they could not hitherto be isolated for direct biochemical analysis; however, a hen cone pigment was long-ago characterized and termed iodopsin. Wald[3] predicted that the cone pigments in general would comprise members of an opsin-like protein family with a retinal chromophore. This prediction has been dramatically confirmed by gene cloning.

Starting with the cloning of the DNA for bovine opsin in 1983[4], molecular cloning studies have been of central importance in research into vision and, by extension, into many other receptor-mediated processes. Several reviews have appeared on aspects of this general field of research (some of these are cited below). The purpose of this article is to summarize, especially for advanced undergraduates and others seeking an introductory review, the unfolding of the opsin gene field in the last 10 years. The story is of considerable intrinsic interest and also highlights the development of techniques and approaches to the study of eukaryotic genes in general

Rhodopsin function and structure in outline

The primary action of light on rhodopsin is to isomerize 11-*cis* retinal to all-*trans* retinal. Retinal is attached to opsin via a protonated Schiff base to a lysine side-chain amino group (Figure 1). Isomerization causes a conformational change in the encompassing opsin and deprotonation of the Schiff base. The altered rhodopsin, called metarhodopsin II or rhodopsin*, activates a signalling cascade whose effect is to generate an impulse by closing ion channels in the rod cell membrane.

Metarhodopsin II activates a G-protein, transducin, by replacement of GDP with GTP in the transducin α-subunit. The activated transducin in turn activates a phosphodiesterase, lowering the cyclic GMP concentration in the rod cell. Depletion of cyclic GMP causes closure of sodium channels, hyperpolarization of the rod cell membrane and electrical discharge. One molecule of metarhodopsin II activates many transducins, each of which activates many phosphodiesterases, greatly amplifying the initial quantal light signal. Metarhodopsin II is turned off somewhat slowly (a time-scale of the order of 0.1 s) by multiple phosphorylations, followed by binding of another protein called arrestin, and the Schiff base linkage to all-*trans* retinal then becomes hydrolysed. Functional rhodopsin is regenerated by binding of a new 11-*cis* retinal.

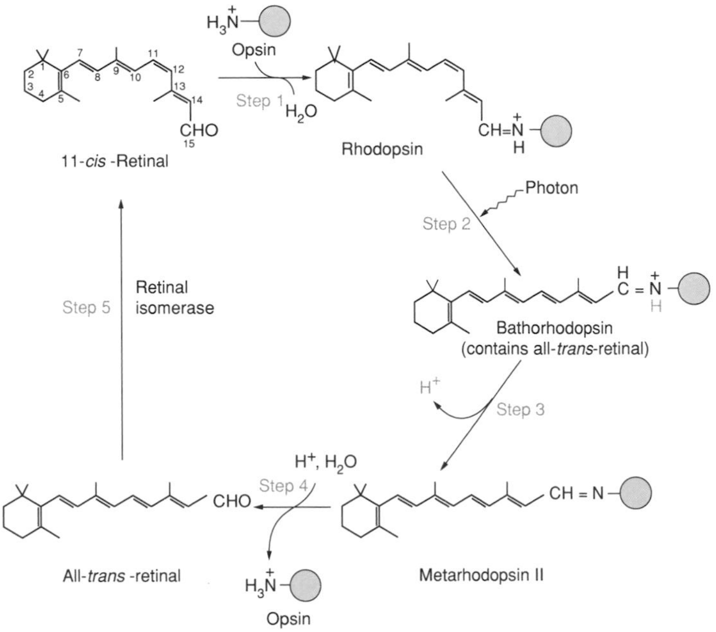

Figure 1. The retinal cycle in rhodopsin
Light isomerizes 11-*cis* retinal in rhodopsin to all-*trans* retinal. The Schiff base linking retinal to lysine is protonated in rhodopsin and in the initial photoproduct, bathorhodopsin, but is deprotonated in metarhodopsin II, the metastable intermediate in signal transduction. For further details, see the text. [Redrawn from Mathews, C.K. & van Holde, K.E. (1990) *Biochemistry*, p. 1052, Benjamin Cummings]

Further understanding of these processes requires knowledge of the structure of rhodopsin, for which the starting point was the determination of the complete amino acid sequence of bovine opsin. This was accomplished both by direct protein sequencing (for summaries of the original work see references 5 and 6) and, as will be further described below, by cDNA sequencing[4]. During protein sequencing the site of retinal attachment was identified as Lys-296.

The polypeptide chain traverses the rod disc membrane seven times (Figure 2). An indication of this arrangement came from the partly comparable protein bacteriorhodopsin. Bacteriorhodopsin is a halobacterial, light-driven, proton pump which also exploits retinal (although, in contrast to rhodopsin, in bacteriorhodopsin retinal is photoisomerized from all-*trans* to 13-*cis*). Structural studies on bacteriorhodopsin revealed seven membrane-spanning helices (for a short review, see reference 7). Although bovine opsin shows no sequence similarity to bacteriorhodopsin, it contains seven hydrophobic tracts, indicating a similar arrangement in the membrane. Five of the seven transmembrane helices of bovine opsin contain proline residues, which must create helix kinks. Enzymic and chemical probing have confirmed the seven transmembrane arrangement, with the *N*-terminal end protruding intradiscally and bearing glycosyl chains, and the *C*-terminal end on the cytoplasmic side[6,8]. The *C*-terminal end contains the protein phosphorylation sites, and also two

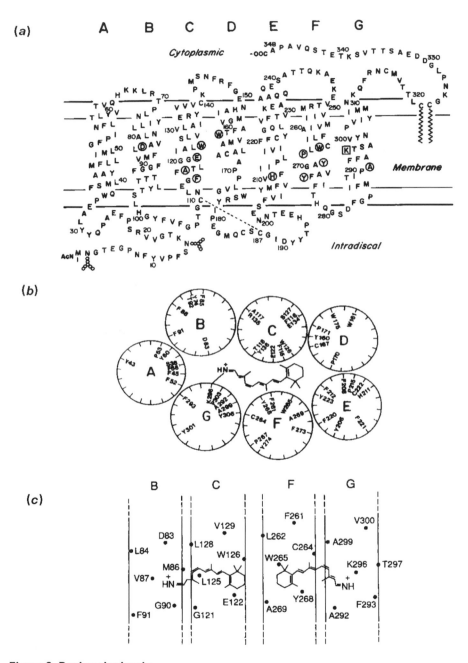

Figure 2. Bovine rhodopsin

(*a*) Secondary structure model. Transmembrane helices are named A to G in the Khorana nomenclature[9,30]; interhelical loops are named AB, CD, EF (cytoplasmic) and BC, DE, FG (intradiscal). Asn-2 and Asn-15 are glycosylated. Lys-296 (boxed) is the retinal attachment site. Cys-321 and Cys-322 are palmitoylated. Serine and threonine residues in the *C*-terminal tail are sites of multiple phosphorylation by rhodopsin kinase. Cys-110, Cys-187 and several of the ringed amino acids are discussed later in the text. (*b*) Probable disposition of the helices around 11-*cis* retinal. (*c*) Location of retinal in relation to the internal aspects of the central parts of helices B, C, F and G. [All sections of this figure are reproduced with permission from reference 30].

palmitoylated cysteine residues that provide further membrane attachment[9]. In both rhodopsin and bacteriorhodopsin there are interhelical loops; in bacteriorhodopsin most of these are minimal in length, in rhodopsin several loops are sizeable and their functions are being elucidated, as will be described later. The model for rhodopsin in the membrane shown in Figure 2 includes a convenient nomenclature[9] for the helices and loops.

Bovine opsin cDNA

By the early 1980s it was clear that the cloning of bovine opsin DNA would break new ground for the study of opsins. The isolation of cloned, full-length cDNA involved an early application of oligonucleotide probing[4]. Only part of the protein sequence was then known. Within the known part a suitable target tract was identified for an oligonucleotide: a minimally degenerate region with two consecutive methionine residues (with unique codons) about 40 residues from the C-terminus.

$$
\begin{array}{llllll}
\text{Ile} & \text{Met} & \text{Met} & \text{Asn} & \text{Lys} \\
\overset{\text{U}}{\underset{\text{A}}{\text{5' AUC}}} & \text{AUG} & \text{AUG} & \text{AA}\overset{\text{U}}{\underset{\text{C}}{}} & \text{AA}\overset{\text{A}}{\underset{\text{G}}{}} & \text{3' Possible} \\
 & & & & & \text{mRNAs} \\
\text{3' TAG} & \text{TAC} & \text{TAC} & \text{TTG} & \text{TTC} & \text{5' DNA}
\end{array}
$$

The oligonucleotide was used as a primer for generating, by reverse transcription, an opsin-enriched radioactive probe. This was used to probe a λ bovine retina cDNA library of some 2×10^5 clones. (For details of the procedure, see reference 4.) Several positive clones were identified, and one of these, called bd20, was essentially full-length opsin cDNA.

The findings from the cDNA were as follows. First, the 348 amino acid sequence deduced from cDNA corresponded exactly with that from the protein, which had meanwhile been completed. Secondly, the cDNA shows no signal sequence; the N-terminal methionine residue (acetylated in the protein) is the translation initiator. Thirdly, although rhodopsin is abundant in the retina, its mRNA is not so abundant, comprising only about 0.5% of retina mRNA. Finally, bovine rhodopsin mRNA contains a long 3′ untranslated region of some 1400 bases. For these last two reasons the attainment of the cloned full-length cDNA was more difficult than might have been anticipated.

In another bovine opsin cDNA study[10] Kuo *et al.* isolated almost full-length clones, fully sequenced the long 3′ untranslated region (which had been only partly sequenced in reference 4), and identified alternative polyadenylation sites. Moreover, they demonstrated mRNAs of sizes corresponding to those specified by the polyadenylation sites and found that their relative abundances were modulated by the state of light- or dark-adaptation of the retina; the shorter mRNA was more abundant in the light-adapted retina, a curious mRNA-processing effect.

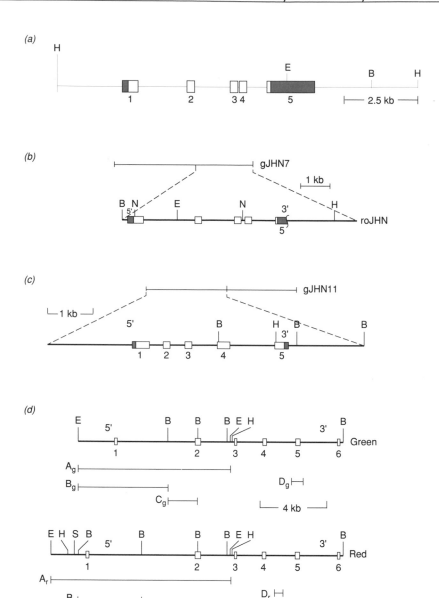

Figure 3. Mammalian opsin genes and their exon–intron structures

(a) The 12 kb HindIII fragment containing the bovine opsin gene. (b) One of four overlapping human Sau3A clones encompassing the human opsin gene, and the subclone roJHN. The bovine and human opsin exon–intron maps are to approximately the same scale. (c) One of four overlapping Sau3A clones encompassing the human blue pigment gene, and the blue gene exon–intron map. (d) The human green pigment and red pigment gene maps. Note that the blue gene map is shown to a larger scale than opsin because, with introns, it encompasses a shorter overall length, and the green and red gene maps are to a smaller scale than opsin because, with introns, they encompass a greater length. Exons (boxes) are numbered as in the text. Shaded areas are non-coding regions. Superscript letters denote restriction sites for BamHI, EcoRI, HindIII, NcoI and SalI. The restriction fragments Ag, Ar; Bg, Br; Cg, Cr; and Dg, Dr differ in length between green and red genes. For further details, see the text. (Adapted with permission from references 4, 11 and 12).

Bovine and human opsin genes

Isolation of cloned cDNA for bovine opsin enabled the cloning of the bovine opsin gene[4], the human opsin gene[11] and the human colour vision genes[12]. The bovine opsin gene was isolated from a λ library that had been constructed from 12 kb, size-fractionated, genomic *Hind*III fragments, following Southern blotting results which suggested that the gene resided within a *Hind*III fragment of that size[4]. The human opsin gene was isolated from a library of DNA[11] from a male (JN) with normal colour vision (see below), after partial digestion with *Sau*3A. Because *Sau*3A is a tetranucleotide cutter, partial digestion is capable of yielding multiple overlapping clones. Four overlapping clones of the human opsin gene were isolated. The gene was wholly within a 7 kb *Bam*HI–*Hind*III fragment (called roJHN), which was subcloned from one of the original λ clones to a plasmid vector (pBR322) for sequencing.

The findings on the bovine[4] and human[11] opsin genes may be summarized together. The protein-coding regions are highly conserved, both genes coding for a protein sequence of 348 residues, with 93.4% conservation of amino acids and no insertions or deletions. Both genes are interrupted by four introns at exactly homologous locations. The second, third and fourth introns are located at sites encoding the C-termini of transmembrane helices D, E and G (for nomenclature, see reference 9) (Figure 2); the first is located internally in the region encoding helix C. The respective intron sizes are also fairly similar (Figure 3*a* and 3*b*) but their sequences are not conserved; intron sequences of eukaryotic nuclear genes generally diverge more rapidly than protein-coding sequences. (The mouse opsin gene analysed later[13] was also found to encode a highly conserved protein, and to contain introns at the same sites and of almost similar sizes to those of the bovine and human genes.) These findings of a highly conserved mammalian rhodopsin set the stage for human colour vision.

Human colour vision

Two years after the characterization of the human opsin gene, the molecular basis of human colour vision was described in two landmark papers[12,14]. The bovine opsin cDNA probe, which had given access to the human opsin gene, also hybridized at lower stringency to several additional clones from the human *Sau*3A library. These additional clones fell into two groups.

The blue pigment gene

One group of four clones encompassed what became identified as the blue pigment gene (hereinafter termed the 'blue' gene). Like opsin, this gene encodes a 348 amino acid protein with seven potential transmembrane helices and four introns. The introns were located by comparison with corresponding cDNA clones, isolated from a human retina library. The blue gene differs from the opsin gene in a number of respects. (i) The deduced amino acid sequence differs very considerably from that of opsin, with only 42% identical and 75% similar amino acids. (ii) The aligned amino acid sequences imply the loss of three amino acids from near the N-terminus of the blue pigment with respect to opsin, and the gain of three amino acids in the C-terminal region. The locations of the retinal-binding lysine residue, several other invariant amino acids (which have been recognized more recently, see later) and the introns are conserved when the sequences are aligned in this way. (iii) The intron sizes are much smaller than in opsin, except for the third intron which is small in opsin but larger in the blue gene (Figure 3c). (iv) The 5' and 3' untranslated regions of mRNA are much shorter than in opsin. Thus the whole gene is more compact than the opsin gene. It was identified as the blue gene by its hybridization to chromosome 7[14]; inherited defects in blue discrimination (which are rare) were already known to be autosomal. In the same study the opsin gene was located to chromosome 3[14].

The green and red pigment genes

Abnormalities in green and red perception are sex-linked and are relatively common in males (about 8% of Caucasian males); the respective loci are tightly linked on the q-arm of the X-chromosome. Molecular genetic analysis of the loci commenced with two further human genomic Sau3A clones that hybridized weakly to bovine opsin cDNA. These proved to be partial clones of two tandemly linked green pigment genes. (This identification will be assumed here for convenience but was established as described in the next section. The green and red pigment genes are hereinafter termed the 'green' and 'red' genes.) Complete recovery of the DNA encompassing the two green genes and the red gene required re-probing of the library by one of the partial green genes and, for the red gene, further rounds of cloning. The picture was completed by isolation of corresponding cDNA clones, and by Southern blotting experiments on several colour-normal males, and is as follows.

The red–green locus comprises a tandemly linked array with the red gene at the 5' end and, in different colour-normal males, either one, two or three green genes. The red and green genes are highly similar in their protein-coding regions, but differ widely from the blue gene and opsin (43% and 40% amino acid identity; 77% and 70% similarity). The red and green genes each contain an extra exon encoding an N-terminal 37 amino acid sequence instead of, and showing no sequence similarity with, the first 21 amino acids of opsin. This first exon is followed by an intron, after which the sequences can be aligned with opsin with no insertions or deletions until the C-terminal tail (see Figure 5, later). The red and green introns 2–5 are at homologous locations to opsin introns 1–4. However, the introns are much larger than in opsin, so that each gene spans 12–13 kb (Figure 3d); the first intron is larger in the red gene than in the green genes. Otherwise, the introns show high sequence similarity between the red and green genes. The findings on intron similarity indicate recent duplication and divergence of the red from the green gene(s). This is consistent with the fact that only Old World primates have red and green cone pigments; New World monkeys have a single, long-wavelength pigment encoded on the X-chromosome.

Red and green abnormalities and the absolute red–green map

As mentioned above, abnormalities in red and green perception are relatively common among Caucasian males. Although not normally life-threatening, the abnormalities do have potential socio-economic consequences; for example, they debar individuals from activities such as aviation piloting. Several classes of abnormality are distinguishable. Dichromats have only a single functioning long-wavelength pigment, either red or green. Anomalous trichromats have two long-wavelength pigments, one of which (the red or green) has an abnormal wavelength maximum between the normal red and green maxima.

The molecular basis of these abnormalities was investigated in some 25 male subjects[14]. The strategy utilized the following facts: the red and green genes are highly similar through most of their lengths, as mentioned above; several restriction sites are conserved between the red and green genes, whereas some sites are unique to red; and the first intron is larger in the red than the green gene. From these considerations it was possible to choose four restriction fragments (A, B, C and D) spanning representative regions of the genes, such that each fragment differed in length between the red and green genes (Figure 3d). By using suitable probes the fragment pairs could be visualized on Southern blots of the DNA of patients and of normal control males. The presence of only a single X-chromosome in males greatly facilitated the interpretation of the blots.

All subjects with normal red perception possessed a large restriction fragment Ar derived from the first two exons of the gene (Figure 3d). A slightly shorter, corresponding fragment Ag spans the first two exons of the green gene. Southern blotting and quantitation of these fragments established the conclusion that different colour-normal males possess one, two or more green genes and only one red gene. A smaller fragment Br derives from within the 5′ region of Ar, i.e. within the 5′ region of the red gene, spanning the first exon (Figure 3d). This fragment was present in Southern blots of all subjects, including those with non-functional or anomalous red perception. Evidently fragment Br remains unaffected by processes generating red dysfunction. Otherwise, the Southern blots showed various permutations of the long and short representatives of the fragment pairs, indicating that anomalies in red–green perception are generated by unequal recombination (or gene conversion) between or within genes in the red–green array.

The constant presence of fragment Br indicated its exclusion from recombination and, hence, location of the red gene at the 5′ end of the array. The abnormal genotypes, diagnosed from the respective combinations of red- and

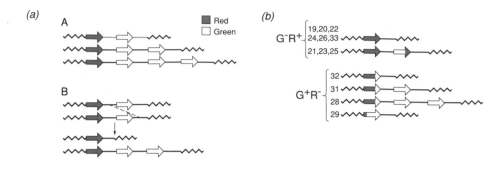

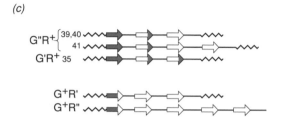

Figure 4. Arrangements of human red and green genes

(a) One red gene and one or more green genes in different colour-normal males, showing how unequal recombination can eliminate or generate different numbers of green genes. (b) Arrangements discovered in dichromats. (c) Arrangements discovered in anomalous trichromats. Each number designates a male subject, with a green (G⁻,G',G") or red (R⁻,R',R") visual defect. For further details, see the text. [Reproduced with permission from reference 12 (a) and reference 14 (b and c).]

green-specific fragments in the Southern blots, include (Figure 4) complete loss of green genes and the generation of various hybrid red–green or green–red genes — with or without one or more normal genes, but always with Br (the 5' end of red) at the 5' end of the array. Hybrid genes with 5' red- and 3' green-coding regions, or vice versa, suggest an explanation for the finding that in anomalous trichromats the abnormal pigment absorbs maximally between the normal red and green maxima (see also section entitled Colour vision: the tuning problem).

Recombination in the red–green array is reminiscent of events in other multigene clusters, such as the α-globin and β-globin clusters, suggesting that the red and green genes are physically closely linked. This was confirmed by probing after digestion with infrequently cutting restriction enzymes and separating the products by a variation of the technique of pulsed-field electrophoresis[15]. The enzyme *Not*I was found to cut on either side of the array, but not within it. This enabled size increments of the array to be determined on DNA from individuals bearing from one to five genes. The DNA size increment for each additional gene was 39 kb. Since each green gene spans approx. 12 kb (see above) the repeat intergenic space is some 27 kb.

By using another enzyme (*Sfi*I) and a probe specific for the 5′ end of the red gene, it was confirmed that red is indeed at the 5′ end of the array; the probe detected a large unique fragment spanning the upstream *Not*I site.

Thus the main features of physical arrangement of the red and green genes inferred earlier[12,14] were confirmed and quantified. The regular spacing between repeats implies that the entire 39 kb unit with its intergenic material is repeated. This facilitates homologous recombination between genes, generating variable numbers of structurally normal green genes. The red–green hybrid genes are produced by intragenic recombination between red and green genes. Presumably the majority of intragenic recombination events occur within the large introns. Thus the high incidence of abnormalities in red and green perception springs from the physical features of the red–green gene cluster.

Drosophila opsins and opsin phylogeny

Flies and many other invertebrates possess compound eyes, which are morphologically very different from vertebrate eyes. Nevertheless, their photoreceptor molecules are also rhodopsins with 11-*cis* retinal as chromophore. Analysis of the molecular genetics of *Drosophila* opsins has illuminated some essential features of the opsin molecule and opsin phylogeny.

The *Drosophila* compound eye consists of many unit structures called ommatidia. Each ommatidium consists of eight photoreceptor cells. Six outer cells, R1–6, express a rhodopsin that absorbs maximally at 480 nm. R7 and R8 occupy the central region of the ommatidia. R8 cells contain an opsin that responds maximally to blue light. R7 cells contain either of two rhodopsins which respond to u.v. light.

Cloning of the genes and cDNAs encoding these opsins commenced with R1–6 opsin[16,17]. The strategy utilized bovine opsin cDNA as a probe. This possessed just sufficient sequence similarity (in limited parts of the sequence) to successfully detect *Drosophila* opsin sequences. Once cloned, the R1–6 opsin gene was used as a probe for the cloning of the R8[18] and R7[19,20] opsins.

The R1–6 opsin[16,17] clearly possessed an overall structure similar to mammalian opsins, with seven transmembrane segments and three regions of substantial amino acid sequence conservation. After the additional opsins had been cloned and sequenced, the *Drosophila*–mammalian sequence alignments were refined[18–20] and an opsin phylogeny was proposed[19]. The conclusions summarized here are from the final two papers[19,20]; see Figures 5 and 6 and Table 1.

The *Drosophila* opsins are longer than mammalian (rod) opsin, with extra amino acid tracts (Figure 5) in the *N*-terminal sequence and cytoplasmic loop EF (for nomenclature, see Figure 2). The regions of highest conservation are in the first cytoplasmic loop AB, external loop DE and helix G, which contains the lysine residue to which retinal is attached. There are 28 invariant amino acids in all of the opsins (Table 1) and several other sites where only

```
hGreen
hRed
hBlue
hRhodopsin
ha β-AR
Dm Rh1
Dm Rh2
Dm Rh3
Dm Rh4
                                                                           *
hGreen        MAQQWSLQRLAGRHPQDSYEDSTQSSIFTYTNSNSTRGPFEGPNYHIAPRWV-          52
hRed          MAQQWSLQRLAGRHPQDSYEDSTQSSIFTYTNSNSTRGPFEGPNYHIAPRWV-          52
hBlue                           MRKMSEEEFYLFKNISSVGPWDGPQYHIAPVWA-           33
hRhodopsin                 MNGTEGPNFYVPFSNATGVVRSPFEYPQYYLAEPWQ-            36
ha β-AR                      MGPPGNDSDFLLTTNGSHVPDHDVTEERDEAWVV             34
Dm Rh1         MESFAVAAAQLGPHFAPLS-NGSVVDKVTPDMAHLISPYWNQFPAMDPIW--          49
Dm Rh2   MERSHLPETPFDLAHSGPRFQAQSSGNGSVLDNVLPDMAHLNVPYWSRFAPMDPMM--       56
Dm Rh3   MESGNVSSSLFGNVSTALRPEARLSAETRLLG--WNVPPEELRHIP-EHWLTYPEPPESM     57
Dm Rh4    MEPLCNASEPPLRPEARSSGNGDLQFLGWNVPPDQIQYIP-EHWLTQLEPPASM          53
```

```
       *         * ***                         *** ***  *        *  *
hG  YHLTSVWMIFVVIASVFTNGLVLAA TMKFKKLR--HPLN WILVNLAVADLAETVIASTISVVN QVYGYFVLGHPMCVLEG  130
hR  YHLTSVWMIFVVTASVFTNGLVLAA TMKFKKLR--HPLN WILVNLAVADLAETVIASTISIVN QVSGYFVLGHPMCVLEG  130
hB  FYLQAAFMGTVFLIGFPLNAMVLVA TLRYKKLR--QPLN YILVNVSFGGFLLCIFSVFPVFVA SCNGYFVFGRHVCALEG  111
hRh FSMLAAYMFLLIVLGFPINFLTLYV TVQHKKLR--TPLN YILLNLAVADLFMVLGGFTSTLYT SLHGYFVFGPTGCNLEG  114
AR   GMAILMSVIVLAIVFGNVLVITA --IAKFERLQTVTN YFITSLACADLVMGLAVVPFGASH ILMKMWNFGNFWCEFWT  110
Rh1 AKILTAYMIMIGMISWCGNGVVIYI FATTKSLR--TPAN LLVINLAISDFGIMITNTPMMGIN LYFETWVLGPHMCDIYA  127
Rh2 SKILGLFTLAIMIISCCGNGVVVYI FGGTKSLR--TPAN LLVLNLAFSDFCMMASQSPVMIIN FYYETWVLGPLWCDIYA  134
Rh3 NYLLGTLYIFFTLMSMLGNGLVIWV FSAAKSLR--TPSN ILVLNLAFCDFMMM-VKTPIFIYN SFHQGYALGHLGCQIFG  134
Rh4 HYMLGVFYIFLFCASTVGNGMVIWI FSTSKSLR--TPSN MFVLNLAVFDLIMC-LKAPIFIYN SFHRGFALGNTWCQIFA  130
```

```
        *         **  **  *** **          *             * *      * *
hG  YTVSLCGITGLWSLAIISWERWMVVC KPFGNVRF-DAKL AIVGIAFSWIWAAVWTAPPIF G-WSRYWPHGLKTSCGPDVF  208
hR  YTVSLCGITGLWSLAIISWERWLVVC KPFGNVRF-DAKL AIVGIAFSWIWSAVWTAPPIF G-WSRYWPHGLKTSCGPDVF  208
hB  FLGTVAGLVTGWSLAFLAFERYIVIC KPFGNFRF-SSKH ALTVVLATWTIGIGVSIPPFF G-WSRFIPEGLQCSCGPDWY  189
hRh FFATLGGEIALWSLVVLAIERYVVVC KPMSNFRF-GENH AIMGVAFTWVMALACAAPPLA G-WSRYIPEGLQCSCGIDYY  192
AR  SIDVLCVTASIETLCVIAVDRYIAIT SPFKYQSLLTKNK ARMVILMVWIVSGLTSFLPIQ MHWYRATHQKAIDCYHKETC  190
Rh1 GLGSAFGCSSINSMCMISLDRYQVIV KGMAGRP-MTIPL ALGKIAYIWFMSSIWCLAPAF G-WSRYVPEGNLTSCGIDYL  205
Rh2 GCGSLFGCVSINSMCMIAFDRYNVIV KGINGTP-MTIKT SIMKILFIWMMAVFWTVMPLI G-WSAYVPEGNLTACSIDYM  212
Rh3 IIGSYTGIAAGATNAFIAYDRFNVIT RPMEGK--MTHGK AIAMIIFIYMYATPWVVACYT ETWGRFVPEGYLTSCTFDYL  212
Rh4 SIGSYSGIGAGMTNAAIGYDRYNVIT KPMNRN--MTFTK AVIMNIIWLYCTPWVVLPLT QFWDRFVPEGYLTSCSFDYL  208
```

```
         *       *   ** **    * *  **           *   *
hG  SGSSYPGVQS YMIVLMVTCCITPLSIIVLCYLQVWLAI-- RAVAKQQKESESTQ------------KAEKEVTR-----  268
hR  SGSSYPGVQS YMIVLMVTCCIIPLAIIMLCYLQVWLAI-- KAVAAQQQESATTQ------------KAEKEVTR-----  268
hB  TVGTKYRSES YTWFLFIFCFIVPLSLICFSTYQLLRAL-- KAVAAQQQESATTQ------------KAEREVSR-----  249
hRh TLKPEVNNES FVIYMFVVHFTIPMIIIFFCYGQLVFTV-- KEAAAQQQESATTQ------------KAEKEVTR-----  252
AR  CDFFTNQAYA --IASSIVSFYVPLVVMMVFVYSR-VFQVAK RQLQKIDKSEGRFHSPNLGQVEQDGR-SGHGLRRSSKFCL  266
Rh1 ERDWNPRSYL --IFYSIFVYYIPLFLICYSYWFIIAAVSA HEKAMREQAKKMNV-KSLRSSEDAEK-SAEGKLAK-----  276
Rh2 TRMWNPRSYL --ITYSLFVYYTPLFLICYSYWFIIAAVAA HEKAMREQAKKMNV-KSLRSSEDCDK-SAEGKLAK-----  283
Rh3 TDNFDTRLFV --ACIFFFSFVCPTTMITYYYSQIVGHVFS HEKALRDQAKKMNV-ESLRSNVDKNKETAEIRIAK-----  284
Rh4 SDNFDTRLFV --GTIFFFSFVCPTLMILYYYSQIVGHVFS HEKALREQAKKMNV-ESLRSNVDKSKETAEIRIAK-----  280
```

```
           *       *  ** **      *                     ▼    *  *** ** **     *    *
hG  ------MVVVMV-LAFCFCWGPYAFFACFAA ANPGYPFHP-- LMAALPAFFAKSATIYNPVIYVFM NRQFRNCILQ---L  336
hR  ------MVVVMI-FAYCVCWGPYTFFACFAA ANPGYAFHP-- LMAALPAYFAKSATIYNPVIYVFM NRQFRNCILQ---L  336
hB  ------MVVVMV-GSFCVCYVPYAAFAMYMV NNRNHGLDL-- RLVTIPSFFSKSACIYNPIIYCFM NKQFQACAMK--MV  318
hRh ------MVIIMV-IAFLICWVPYASVAFYIF THQGSNFGP-- IFMTIPAFFAKSAAIYNPVIYIMM NKQFRNCMLT-TIC  322
AR  KEHKAL KTLGIIMGTFTLCWLPFFIVNIVHV --IQDNLIPKE VYILLNWLGYVNSA-FNPLIYCRS -PDFRIAFQELLCL  342
Rh1 ------VALVTITLWFMA-WTPYLVINCMGL F-KFEGLTP-- LNTIWGACFAKSAACYNPIVYGIS HPKYRLALKE-KCP  345
Rh2 ------VALTTISLWFMA-WTPYLVICYFGL F-KIDGLTP-- LTTIWGATFAKTSAVYNPIVYGIS HPKYRIVLKE-KCP  352
Rh3 ------AAITICFLFFCS-WTPYGVMSLIGA FGDKTLLTP-- GATMIPACACKMVACIDPFVYAIS HPRYRMELQK-RCP  354
Rh4 ------AAITICFLFFVS-WTPYGVMSLIGA FGDKSLLTP-- GATMIPACTCKLVACIDPFVYAIS HPRYRLELQK-RCP  350
```

```
          **        *
hG  FGKKVDDGSEL-SSASKTEVSSVSSVSPA                                        364
hR  FGKKVDDGSEL-SSASKTEVSSVSSVSPA                                        364
hB  CGKAMTDESDTCSS-QKTEVSTVSSTQVGPN                                      348
hRh CGKNPLGDDEASATVSKTETSQVAPA                                           348
AR  RRSSSKAYGNGYSSNGNGKTDYMGEASGCQLGQEKESERLCEDPPGTESFVNCQGTVPSLSLDSQGRNCSTNDSPL  418
Rh1 CCVFGKVDD-GKSSDAQSQ-ATASEAESKA                                       373
Rh2 MCVFGNTDE-PKPDAPASDTETTSEADSKA                                       381
Rh3 -WLALNEKA-PESSAVAS-TSTTQEPQQTTAA                                     383
Rh4 -WLGVNEKSGEISSAQS---TTTQEQQQTTAA                                     378
```

Figure 5. Aligned sequences of the human green, red and blue opsins, human rhodopsin, the hamster β-adrenergic receptor and the *Drosophila* opsins 1–4

Drosophila opsin 1 is that described in references 16 and 17; opsins 2, 3 and 4 are those described in references 18, 19 and 20, respectively. Boxed areas are the transmembrane segments as deduced in reference 19 (the boundaries in Figure 2, deduced more recently, differ slightly in a few instances, see the text). Asterisks indicate homologies between the β-adrenergic receptor and at least four opsins. The arrow indicates the lysine residue to which retinal is attached; valine occurs at this site in the β-adrenergic receptor. (Reproduced with permission from reference 19.)

Table 1. Amino acids conserved between mammalian and *Drosophila* opsins

| Residue | Location | | | Notes |
	Intradiscal	Transmembrane	Cytoplasmic	
Pro-23	*N*-terminal tail			a
Asn-55		Helix A		a
Lys-66			Loop AB	a
Leu-68			Loop AB	
Arg-69			Loop AB	a
Pro-71			Loop AB	
Asn-73			Loop AB	a
Asn-78		Helix B		
Gly-106	Loop BC			a
Cys-110	Loop BC			a,b
Gly-121		Helix C		
Arg-135		Helix C		a
Val-138			Loop CD	
Trp-175		Helix D		a,c
Pro-180	Loop DE			
Gly-182	Loop DE			
Cys-187	Loop DE			b
Asp-190	Loop DE			a
Pro-215		Helix E		
Ile-219		Helix E		
Ala-234			Loop EF	
Ala-246			Loop EF	
Glu-247			Loop EF	
Pro-267		Helix F		a
Tyr-268		Helix F		
Lys-296		Helix G		d
Pro-303		Helix G		a
Tyr-306		Helix G		a

These amino acids are conserved between all of the following: bovine and human rhodopsins, the human blue, green and red opsins, and all four *Drosophila* opsins. Numbering is according to mammalian rhodopsin (see Figure 2). For the corresponding numbering in the human colour opsins and the *Drosophila* opsins, see the aligned sequences in Figure 5. Notes: (a) these residue are also identical at the corresponding locations in the hamster and human β-adrenergic receptors[22,23]; (b) these two cysteines form a disulphide bond — in the β-adrenergic receptor, the second cysteine corresponding to Cys-187 in mammalian rhodopsin is one place to the left in the aligned sequences (Figure 5); (c) this tryptophan is close to the boundary between helix D and loop DE (Figure 2); (d) retinal attachment site.

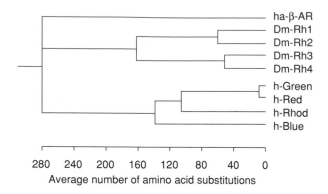

Figure 6. Phylogeny of visual pigments and the β-adrenergic receptor, constructed
on the basis of the principle of minimum mutation distances (parsimony principle)
Given that the human and bovine rhodopsins differ by only 23 amino acids, after approx
60 Myear of mammalian evolution, it can be inferred from this tree that the human blue,
rhodopsin and long-wavelength genes are from lineages that diverged early in vertebrate evolution. For further details of the phylogenetic tree construction, see reference 19. (Reproduced
with permission from reference 19.)

conservative substitutions occur. Possible functions were proposed for some
of the conserved sites, including a disulphide bond between two cysteine
residues in extracellular loops BC and DE. Direct evidence on the functions of
these and other residues has come from further lines of investigation, as
described later.

The R1–6 *Drosophila* opsin contains four introns, three of which are in
approximately similar locations to those in mammalian rhodopsin. However,
in the other *Drosophila* opsins the number of introns ranges from zero to
three, and the locations of some of these differ from those in the R1–6 opsin[20].
Thus the constancy of four intron sites in mammalian (vertebrate?) opsins
does not hold for all opsins.

From the *Drosophila* and human sequences, Zuker *et al.*[19] constructed an
opsin phylogeny (Figure 6). The essential features of human opsin phylogeny
had also been deduced previously from the colour opsin data[12]. The combined
data lead to the inference that about 10^9 years ago an ancestral opsin gene
began to diverge between invertebrate and future vertebrate lineages. Gene
duplications and further divergences within both lineages have given rise to
present day opsins. In particular, the genes that gave rise to the primate blue
opsin, rhodopsin and the precursor to the red–green opsins have diverged
over 5×10^8 years from the beginning of vertebrate evolution. Interestingly,
vertebrate α- and β-globin divergence commenced at about the same time[21].
Ancient opsin duplications and divergences also occurred within the lineage
from which *Drosophila* evolved. Finally, opsins show significant homology to
the β-adrenergic receptors (Figures 5 and 6), as will now be discussed briefly.

The seven transmembrane helix superfamily of receptors

The β-adrenergic receptors are among many membrane receptors which, like rhodopsin, mediate signal transduction by activation of a G-protein. Cloning of the gene and cDNA for the hamster β_2-adrenergic receptor[22] followed by the cDNA for the human β_2-adrenergic receptor[23] showed that this protein too contains seven hydrophobic transmembrane regions, which form the binding pocket for the adrenergic agonist. In both instances, the cDNA isolation was technically noteworthy for the low abundance of positive clones, about two per 10^6 recombinants in the respective libraries. Since then numerous further G-protein-coupled receptors have been cloned and shown to possess a seven transmembrane structure. These include many different neurotransmitter receptors, peptide receptors and a large family of putative olfactory[24] receptors. (For general reviews, see references 25 and 26.) The receptor proteins range in length from about 310 amino acids to more than 500 amino acids, with the greatest length variation in the N-terminal extracellular and C-terminal cytoplasmic tails and cytoplasmic loop EF. Several amino acid residues are conserved through most members of the superfamily, including the two cysteine residues that form the extracellular disulphide bond. (For some representative aligned sequences, see reference 26.)

Recognition that rhodopsins are members of a larger superfamily of G-coupled receptor proteins has brought broad significance to efforts to understand the mechanisms of rhodopsin function.

Exploring the details of structure and function of opsins

In the last few years opsin gene research has intensified enormously. Three approaches are mutually addressing the details of opsin structure–function relationships, and a brief overview of these now follows.

Site-directed mutagenesis of opsin

Site-directed mutagenesis enables the identification of amino acid residues that are critical to function. To expedite this approach, Khorana and co-workers synthesized a gene encoding bovine opsin[27] and developed an opsin expression system in cultured, transfected monkey kidney cells[28]. The synthetic gene was constructed in stages from 72 individual oligonucleotides. It differs from the original cDNA clone bd20[4] in containing 28 unique restriction sites that are on average 60 bp apart, enabling the substitution of any desired region with a mutated oligonucleotide. The synthetic gene has no introns but is expressed from a plasmid which contains several features that permit efficient expression of cDNA in mammalian cells (see reference 28 and references 8–10 therein). In this system opsin is expressed to 0.3% of the total cell protein, it binds exogenously added 11-*cis* retinal to yield the correct rhodopsin absorption spectrum, and the rhodopsin recovered from the cells (by immunoaffinity) activates transducin in a light-dependent manner. The rhodopsin is glycosylated and is oriented in the cell membranes with the N-terminus outside, corresponding to its intradiscal location in rods.

The experimental system has allowed the dissection of structure and function in rhodopsin by numerous mutagenic studies, revealing essential features in the external (intradiscal) and cytoplasmic faces and in the transmembrane helices.

On the intradiscal face, the disulphide bond Cys-110–Cis-187, whose presence had been inferred from sequence and protein chemistry data, has been shown by mutagenesis to be essential for formation of functional rhodopsin. Moreover, deletion mutants in each of the three external loops and the N-terminal tail inhibit chromophore binding and rhodopsin transport from the endoplasmic reticulum to the plasma membrane. It is, therefore, inferred that the intradiscal loops and N-terminal tail interact co-operatively to form a folded intradiscal structure which is stabilized by the disulphide bond[9].

On the cytoplasmic face, several deletion mutants in loops CD and EF block transducin activation, and amino acid substitutions in loop EF affect activation to various extents, the results indicating that the C-terminal part of loop EF, in particular, is essential. Also essential is a charged pair, Glu-134–Arg-135, at the cytoplasmic end of helix C, which is conserved in all G-coupled receptors. Reversal of the charged pair abolishes binding, and substitution of the charged amino acids by neutral amino acids seriously affects activation. (For original references to these and the intradiscal face findings, see reference 9.)

For photochemistry of rhodopsin, the transmembrane helices are crucial, for they interact with retinal, feel its isomerization and transmit the signal to the cytoplasmic face for transducin activation. By use of a photoactivatable analogue of 11-cis retinal[29], and further by site-directed mutagenesis[30], several amino acids were identified as contact sites for retinal. These include Phe-115, Ala-117, Glu-122, Trp-126 and Ser-127 in helix C, and Trp-265 and Tyr-268 in helix F. In particular, the bulky Trp-265 is in close contact with the β-ionone ring of retinal: mutation of this residue to alanine severely inhibits reconstitution of rhodopsin from opsin and retinal; mutation to tyrosine or phenylalanine gives abnormal bleaching and reduces V_{max} for transducin activation. Evidently Trp-265 plays a critical role in rhodopsin assembly and in the structural changes involved in bleaching and signal activation.

Glu-113 has been identified by mutagenesis as the counterion to the retinal-lysine protonated Schiff base[31–33]. Mutation to glutamine causes a massive (pH-dependent) shift in spectral absorption from 500 to 380 nm, consistent with the loss of Schiff base protonation at neutral pH. By contrast, mutation of Glu-122 to glutamine or aspartate results in a much smaller blue shift of 20–25 nm[31]. Mutation of other carboxylate amino acids has little or no effect on spectral absorption maxima[31–33]. Thus Glu-113 must be critically juxtaposed to the retinal attachment site and is included in helix C (Figure 2a). It follows that Glu-134 and Arg-135 are near the cytoplasmic end of helix C, consistent with their interaction with transducin. The involvement of Cys-110 in the external disulphide bond means that the transmembrane boundaries of helix C are now fairly well-defined (Figure 2*a*).

Lys-296, the retinal attachment site, can be mutated to glycine or alanine, and functional rhodopsin can be reconstituted by binding of 11-*cis* retinal linked by a Schiff base to an *N*-alkylamine of comparable length to the lysine side-chain[34]. In the absence of chromophore, the mutated rhodopsin activates transducin constitutively; likewise, mutation at the counterion site Glu-113 causes constitutive activation of transducin in the absence of chromophore[35]. A possible interpretation is that the salt bridge, between the protonated Schiff base at Lys-296 and the counterion Glu-113, constrains rhodopsin in the inactive conformation by masking the transducin activation site. Even more interestingly, mutation at Lys-296 to glutamate causes constitutive transducin activation even in the presence of 11-*cis* retinal[35]. Such a mutant occurs in a family with autosomal dominant retinitis pigmentosa.

Autosomal dominant retinitis pigmentosa and molecular pathology of opsin

Retinitis pigmentosa is a heterogeneous group of inherited degenerative disorders, primarily of rods, affecting about one in 4000 individuals. Several genetically distinct categories exist: autosomal dominant and recessive, and X-linked. Among the clinical features are night-blindness that develops in childhood, leading to progressive decrease of the visual fields and eventually 'tunnel vision' in which only the cone-dominated fovea remains functional.

After the mapping of human opsin to the long (q) arm of chromosome 3[14], it was found that a large family with an autosomal dominant form of retinitis pigmentosa (ADRP) bore a distinctive form of polymorphic DNA marker on 3q[36], suggesting opsin as a possible candidate gene for the disease. Soon afterwards an opsin genomic sequencing study (by polymerase chain reaction amplification of each opsin exon of the DNA of affected family members) revealed a Pro-23 → His mutation[37]. Proline is conserved at this position in all opsins (Table 1) and some other G-coupled receptors. Evidently it plays an essential role in the external (intradiscal) domain of rhodopsin (as well as in other G-coupled receptors), the mutation somehow causing retinal degeneration in the family.

Analysis of opsin DNA from many other ADRP families has revealed many further mutations (Table 2)[38–41], and some generalizations are emerging. About 25–30% of ADRP families reveal mutant opsins[39,40]. In Caucasian Americans, the Pro-23 → His mutant is the commonest one, but this mutation has not been found in Europeans, suggesting that the mutation occurred in an early American immigrant family and gave rise to a founder effect. Other mutations occur sporadically or in small pedigrees[39,40]. At least two mutants correspond to positions that had been investigated by site-directed mutagenesis: Arg-135 (transducin activation) and Lys-296 (retinal attachment).

Table 2. Some human opsin mutants in familial retinopathies

| | Location | | | References |
Mutation	Intradiscal	Transmembrane	Cytoplasmic	and notes
Thr-17 → Met	N-terminal tail			39
Pro-23 → His	N-terminal tail			37,39 (a)
Phe-45 → Leu		Helix A		39
Pro-53 → Arg		Helix A		40
Thr-58 → Arg		Helix A		40
68–71 deletion			Loop AB	38 (a)
Val-87 → Asp		Helix B		39
Gly-89 → Asp		Helix B		39
Gly-106 → Trp	Loop BC			39 (a)
Gly-106 → Arg	Loop BC			40 (a)
Arg-135 → Leu		Helix C		39 (a)
Arg-135 → Trp		Helix C		39 (a)
Tyr-178 → Lys	Loop DE			39
Asp-190 → Asn	Loop DE			38 (a)
Asp-190 → Gly	Loop DE			39 (a)
Met-207 → Arg		Helix E		41
His-211 → Pro		Helix E		38
Ile-255 deletion		Helix F		40
Ala-292 → Glu		Helix G		42 (b)
Lys-296 → Glu		Helix G		38 (a)
Gln-344 → stop			C-terminal tail	39
Pro-347 → Leu			C-terminal tail	39,40

Notes: (a) Pro-23, Leu-68, Arg-69, Pro-71, Gly-106, Arg-135, Asp-190 and Lys-296 are conserved in all of the human and *Drosophila* opsins in Table 1; (b) from a family with congenital stationary night blindness. All the other mutations are from families with ADRP.

Evidently a variety of natural mutants affecting the intradiscal, transmembrane and cytoplasmic domains of opsin can give rise to the ADRP phenotype. Different mutants result in different degrees of severity of the disease, and presumably cause retina degeneration in different ways. For example, the mutant Lys-296 → Glu, through being constitutive for transducin activation, may 'wear out' the rod cell's metabolic machinery[35]. In other mutants, the disease phenotype may be connected with failure to assemble functional rhodopsin. A less severe phenotype, 'stationary' (non-progressive) night blindness also exists, and a family bearing this condition has recently been shown to carry a mutation Ala-292 → Glu in helix G[42].

Neutral amino acid substitutions in human rhodopsin are evidently very rare. The study by Sung *et al.*[39] encompassed 161 unrelated ADRP patients and 118 control subjects with normal vision. Only 39 of the ADRP patients revealed opsin mutations, and no mutations unrelated to ADRP were discovered.

Among ADRP patients for whom no opsin gene mutations were found it is possible that mutations may exist in introns or control regions in some patients. However, some forms of ADRP arise from other genes, including, as recently discovered, that for the intradiscal protein peripherin[43].

It should, finally, be mentioned that a family has been described with a non-functional cone pigment in which one of the disulphide cysteine residues, Cys-203, corresponding to Cys-187 in opsin, is mutated to arginine[44].

Colour vision: the tuning problem

The directed or spontaneous mutants summarized in the preceding two sections disrupt rhodopsin function to varying extents. But nature has also carried out successful protein engineering to produce the different colour opsins. A full understanding of the determinants of photopigment spectral sensitivity, the 'tuning' problem, is perhaps the greatest fundamental challenge in opsin research. Nathans[45,46], has reviewed much of the background.

Retinal in the form of a protonated Schiff base in methanol, without opsin, absorbs maximally at 440 nm. Therefore, in rhodopsin (absorption max. 500 nm), a red shift of 60 nm is imposed. This protein-imposed spectral shift is called the 'opsin shift'. In the green and red receptors the opsin shifts are approximately 90 nm and 120 nm to the red, respectively. In the blue receptor a blue shift of somewhat less than 20 nm is imposed.

Experiments with model compounds show that increased delocalization of electrons in conjugated polyene systems causes a red shift, i.e. decreases the energy of photoexcitation. A special class of red shift, distinct from the opsin shift, occurs in many freshwater fish which contain photopigments in which 11-*cis* dehydroretinal, with an extra double bond, replaces 11-*cis* retinal. The resulting 'porphyropsins' have longer absorption maxima values than the corresponding rhodopsins. However, modification of the retinal chromophore is neither a widespread nor a finely variable key to adjusting absorption maxima. Instead, the opsins themselves generally mediate tuning.

One theoretical model sought to explain tuning in terms of the distribution of charged amino acids in the vicinity of the chromophore[47]. The sequences of the human opsins seemed to provide initial support for a point charge model. The net charge of amino acids in the membrane segments is: blue opsin, + 1; rhodopsin, 0; green and red opsins, each -1[12]. However, attempts to mutate the (bovine) rhodopsin gene at sites where charge differences occur between rhodopsin and the colour opsins yielded only small spectral shifts[32,46]. Moreover, the red and green pigments do not differ in their charged amino acids, indicating that at least some spectral tuning can be mediated by other amino acid differences.

It will be recalled (see section entitled Human colour vision) that humans and Old World monkeys possess distinct red and green receptors, whereas New World monkeys have a single, long-wavelength receptor encoded on the X-chromosome. Nevertheless, there is within-species polymorphism and

Table 3. (i) Three residues affecting long-wavelength sensitivity in New World monkeys[48]

Residue	Absorption max. (nm)	Difference (nm)	Helix	Exon
Ala-180 → Ser	556 → 562	6	D	3
Phe-277 → Tyr	547 → 556	9	F	5
Ala-285 → Thr	541 → 556	15	F	5

Predicted combined total: 30. Residues are numbered as in the red–green numbering scheme (Figure 5). The corresponding sites in human rhodopsin are Ala-164, Phe-261 and Ala-269, each of which is fairly centrally located in the transmembrane helices (Figure 2).

(ii) Absorption maxima (nm) of human cones, human colour pigments and recombinant red–green pigments

| | Cones | Pigments | | |
	Ref. 2	Ref. 53	Refs. 50,51	Notes
Blue	419.0 ± 3.6	424	426	a
Green	527.8 ± 1.8	530	529.7	
	533.7 ± 2.1			
Red	554.2 ± 2.3		552.4	b
	563.2 ± 3.1	560	556.7	c
Recombinants				
R2G3			529.5	d
R3G4 Ala-180			529.0	
R3G4 Ser-180			533.3	
R4G5 Ala-180			531.6	
R4G5 Ser-180			536.0	
G2R3 Ala-180			549.6	
G2R3 Ser-180			553.0	
G3R4			548.8	
G4R5			544.8	

Notes: (a) all pigments were expressed in cultured cells and combined with 11-*cis* retinal for spectroscopy as described[50,51,53]; (b) the red pigment with Ala-180; (c) red pigment with Ser-180; (d) the recombinants are cDNAs with exon sequences of the red and green genes. Thus R2G3 means red exons 1 and 2 plus green exons 3–6, etc. Constructs containing the Ala-180 and Ser-180 variants of red exon 3 were generated.

between-species variation in long-wavelength spectral sensitivity among related species of South American monkeys. Neitz et al.[48] searched for amino acid differences underlying the different spectral sensitivites, using exon-specific polymerase chain reaction and sequencing (as described above for ADRP).

Their findings, together with the human red–green sequence data[12], identified three sites that appeared to account for most if not all of the 30 nm spectral sensitivity differences between human red and green pigments. At each site, replacement of a non-polar amino acid with a hydroxyl-bearing amino acid in one of the transmembrane helices produces a red shift [Table 3(i)].

The human red gene had previously been found to contain a polymorphic Ala-Ser site at one of these three sites, namely residue 180[12]. This polymorphism has recently been shown to underly a 4–5 nm difference in red sensitivity among 'normal' males[49], and of absorbance by the respective red pigments[50] [Table 3(ii)]. Merbs and Nathans[51] have further investigated the determinants of spectral absorption maxima among red–green pigments by constructing hybrid red–green cDNAs and expressing them in cultured cells. Exon 5 was the major determinant of wavelength sensitivity. This exon contains the two other sites where Neitz et al.[48] had shown that tuning occurs (Table 3). Moreover, amino acid changes elsewhere in the molecule modulate the effects of the two critical amino acids in exon 5. (For a short review of red–green tuning, see reference 52.)

What determines the full range of spectral differences between the blue pigment, rhodopsin and the long-wavelength pigments? The evidence summarized above suggests that single substitutions among functional genes in natural populations modulate spectral sensitivity by small amounts — of no more than, and usually less than, 15 nm each. What is the minimum number of substitutions that might shift the spectral sensitivity by more than 100 nm from long wavelength (red–green) to blue (or vice versa)? Do a limited number of effector substitutions among many neutral ones hold the key (recall that neutral mutations in human rhodopsin are rare), or is more complex restructuring of the chromophore microenvironment required to evoke the required electronic properties in 11-*cis* retinal for the different visual pigments?

To enable a systemic approach to spectral tuning, Oprian and colleagues[53] have constructed synthetic cDNAs for the human blue, green and red genes using the same principles as those described above for the (bovine) rhodopsin gene[27]. The pigments expressed from the genes indeed yield the correct spectra of the three human colour receptors, as also reported for the natural cDNAs by Merbs and Nathans[50], [Table 3(ii)]. Other relevant developments are the isolation and molecular characterization of an increasing array of photopigments and their genes from other species[54–57]. Very recently, a u.v.-absorbing rhodopsin has been identified in zebra fish and its cDNA sequenced[58]. Despite having a $\lambda_{max.}$ of 360 nm, it contains Glu-113, in common with bovine rhodopsin and other vertebrate opsins. However, the normal effect of this counterion to the retinylidene Schiff base (see above) may be negated in this opsin by the presence of a charged amino acid, Lys-126, which is unusual in helix C[58]. Thus the molecular basis of spectral tuning, underlying the historic and fascinating problem of colour vision, is entering a new era of rapid and exciting experimental attack.

Concluding comments

The findings summarized in this article, to which the molecular genetic approach has contributed centrally, lead to a view in which all opsins share common features, many of which are also shared by other G-protein-coupled receptors, with individual proteins representing variants on the common theme. The principal common features are a structured external domain, a lig-and-binding transmembrane domain, encompassing retinal in opsins, and a cytoplasmic domain that triggers a G-protein. In opsins, triggering is initiated by isomerization of retinal; this causes conformational change in the encompassing transmembrane domain. The external domain, stabilized by the disulphide bond, is thought to operate as a hinge whereby the transmembrane movements are transmitted to the cytoplasmic face, opening up the transducin activation site. Many mutations are disruptive to opsin function, as revealed by the clinical condition ADRP. The most characteristic differences between opsins are their spectral sensitivities, and the molecular basis of tuning is entering into a phase of intensive experimentation. Further insight into the atomic details of structure–function relationships are likely to come as and when 3-dimensional structural techniques become applicable to opsins.

Of great interest, but not addressed in this article, are the chromosomal and transcriptional mechanisms whereby rhodopsin and the colour opsins are each uniquely expressed in their respective photoreceptor cells.

This article is dedicated to the memory of the late Professor R.A. Morton of Liverpool University, a pioneer of retinal chromophore research. I thank G.A.J. Pitt for helpful comments, Beryl Foulkes for careful typing of this manuscript, and the Royal National Institute for the Blind for support of retina research in my laboratory.

References

1. Wald, G. & Brown, P.K. (1958) Human rhodopsin. *Science* **127**, 222–226
2. Dartnall, H.J.A., Bowmaker, J.K. & Mollon, J.D. (1983) Human visual pigments: microspectrophotometric results from the eyes of seven persons. *Proc. R. Soc. London B.* **220**, 115–130
3. Wald, G. (1968). The molecular basis of visual excitation. *Nature (London)* **219**, 800–807
4. Nathans, J. & Hogness, D.S. (1983) Isolation, sequence analysis and intron–exon arrangement of the gene encoding bovine rhodopsin. *Cell* **34**, 807–814
5. Ovchinnikov, Y.A. (1982) Rhodopsin and bacteriorhodopsin: structure–function relationships. *FEBS Lett.* **148**, 179–191
6. Dratz, E.A. & Hargrave, P.A. (1983) The structure of rhodopsin and the rod outer segment disk membrane. *Trends Biochem. Sci.* **8**, 128–131
7. Khorana, H.G. (1988) Bacteriorhodopsin, a membrane protein that uses light to translocate protons. *J. Biol. Chem.* **263**, 7439–7442
8. Findlay, J.B.C. & Pappin, D.J.C. (1986) The opsin family of proteins. *Biochem. J.* **238**, 625–642
9. Khorana, H.G. (1992) Rhodopsin, photoreceptor of the rod cell. *J. Biol. Chem.* **267**, 1–4
10. Kuo, C.H., Yamagata, K., Moyzis, R.K., Bitensky, M.W. & Miki, N. (1986) Multiple opsin mRNA species in bovine retina. *Mol. Brain. Res.* **1**, 251–260

11. Nathans, J. & Hogness, D.S. (1984) Isolation and nucleotide sequence analysis of the gene encoding human rhodopsin. *Proc. Natl. Acad. Sci. U.S.A.* **81**, 4851–4855

12. Nathans, J., Thomas, D. & Hogness, D.S. (1986) Molecular genetics of human color vision: the genes encoding the blue, green and red pigments. *Science* **232**, 193–202

13. Baehr, W., Falk, J.D., Bugra, K., Triantafyllos, J.T. & McGinnis, J.F. (1988) Isolation and analysis of the mouse opsin gene. *FEBS Lett.* **238**, 253–256

14. Nathans, J., Piantanida, T.P., Eddy, R.L., Shows, T.B. & Hogness, D.S. (1986) Molecular genetics of inherited variation in human color vision. *Science* **232**, 203–210

15. Vollrath, D., Nathans, J. & Davis, R.W. (1988) Tandem array of human visual pigment genes at Xq28. *Science* **240**, 1669–1671

16. O'Tousa, J.E., Baehr, W., Martin, R.L., Hirsh, J., Pak, W.L. & Applebury, M.E. (1985) The Drosophila ninaE gene encodes an opsin. *Cell* **40**, 839–850

17. Zuker, C.S., Cowman, A.F. & Rubin, G.M. (1985) Isolation and structure of a rhodopsin gene from *D. melanogaster*. *Cell* **40**, 851–858

18. Cowman, A.F., Zuker, C.S. & Rubin, G.M. (1986) An opsin gene expressed in only one photoreceptor cell type of the *Drosophila* eye. *Cell* **44**, 705–710

19. Zuker, C.S., Montell, C., Jones, K., Laverty, T. & Rubin, G.M. (1987) A rhodopsin gene expressed in photoreceptor cell R7 of the *Drosophila* eye: homologies with other signal transducing molecules. *J. Neurosci.* **7**, 1550–1557

20. Montell, C., Jones, K., Zuker, C. & Rubin, G. (1987) A second opsin gene expressed in the ultraviolet-sensitive R7 photoreceptor cells of *Drosophila melanogaster*. *J. Neurosci.* **7**, 1558–1566

21. Jeffreys, A.J., Harris, S., Barrie, P.A., Wood, D., Blanchetot, A. & Adams, S.M. (1983) Evolution of gene families: the globin genes in *Evolution from Molecules to Men* (Bendall, D.S., ed.) pp. 175–195, Cambridge University Press, Cambridge

22. Dixon, R.A.F., Kobilka, B.K., Strader, D.J., Benovic, J.L., Dohlman, H.G., Frielle, T., Bolanowski, M.A., Bennett, C.D., Rands, E., Diehl, R.E., *et al.* (1986) Cloning of the gene and cDNA for mammalian β-adrenergic receptor and homology with rhodopsin. *Nature (London)* **321**, 75–79

23. Kobilka, B.K., Dixon, R.A.F., Frielle, T., Dohlman, H.G., Bolanowski, M.A., Sigal, I.S., Yang-Feng, T.L., Francke, U., Caron, M.G. & Lefkowitz, R.J. (1987) cDNA for the human β_2-adrenergic receptor: a protein with multiple membrane-spanning domains and encoded by a gene whose chromosomal location is shared by that of the receptor for platelet-derived growth factor. *Proc. Natl. Acad. Sci. U.S.A.* **84**, 46–50

24. Buck, L. & Axel, R. (1991) A novel multigene family may encode odorant receptors: a molecular basis for odor recognition. *Cell* **65**, 175–187

25. Dohlman, H.G., Thorner, J., Caron, M.G. & Lefkowitz, R.J. (1991) Model systems for the study of seven-transmembrane segment receptors. *Annu. Rev. Biochem.* **60**, 653–688

26. Savarese, T.M. & Fraser, C.M. (1992) *In vitro* mutagenesis and the search for structure–function relationships among G-protein-coupled receptors. *Biochem. J.* **283**, 1–19

27. Ferretti, L., Karnik, S.S., Khorana, H.G., Nassal, M. & Oprian, D.D. (1986) Total synthesis of a gene for bovine rhodopsin. *Proc. Natl. Acad. Sci. U.S.A.* **83**, 599–603

28. Oprian, D.D., Mollday, R.S., Kaufman, R.J. and Khorana, H.G. (1987) Expression of a synthetic bovine rhodopsin gene in monkey kidney cells. *Proc. Natl. Acad. Sci. U.S.A.* **84**, 8874–8878

29. Nakayama, T.A. & Khorana, H.G. (1990) Orientation of retinal in bovine rhodopsin determined by cross-linking using a photoactivatable analogue of 11-*cis*-retinal. *J. Biol. Chem.* **265**, 15762–15769

30. Nakayama, T.A. and Khorana, H.G. (1991) Mapping of the amino acids in membrane-embedded helices that interact with the retinal chromophore in bovine rhodopsin. *J. Biol. Chem.* **266**, 4269–4275

31. Sakmar, T.P., Franke, R.R. & Khorana, H.G. (1989) Glutamic acid-113 serves as the retinylidene Schiff base counterion in bovine rhodopsin. *Proc. Natl. Acad. Sci. U.S.A.* **86**, 8309–8313

32. Zhukovski, E.A. & Oprian, D.D. (1989) Effect of carboxylic acid sidechains on the absorption maximum of visual pigments. *Science* **246**, 928–930

33. Nathans, J. (1990) Determinants of visual pigment absorbance: identification of the retinylidene Schiff's base counterion in bovine rhodopsin. *Biochemistry* **29**, 9746–9752

34. Zhukovsky, E.A., Robinson, P.R. & Oprian, D.D. (1991) Transducin activation by rhodopsin without a covalent bond to the 11-*cis*-retinal chromophore. *Science* **251**, 558–560

35. Robinson, P.R., Cohen, G.B., Zhukovsky, E.A. & Oprian, D.D. (1992) Constitutively active mutants of rhodopsin. *Neuron* **9**, 719–725

36. McWilliam, P., Farrar, G.J., Kenna, P., Bradley, D.G., Humphries, M.M., Sharp, E.M., McConnell, D.J., Lawler, M., Sheils, D., Ryan, C., *et al.* (1989) Autosomal dominant retinitis pigmentosa (ADRP): localization of an ADRP gene to the long arm of chromosome 3. *Genomics* **5**, 619–622

37. Dryja, P., McGee, T.L., Reichei, E., Hahn, L.B., Cowley, G.S., Yandell, D.W., Sandberg, M.A. & Berson, E.L. (1990) A point mutation of the rhodopsin gene in one form of retinitis pigmentosa. *Nature (London)* **343**, 364–366

38. Keen, T.J., Inglehearn, C.F., Lester, D.H., Bashir, R., Jay, M., Bird, A.C., Jay, B. & Bhattacharya, S.S. (1991) Autosomal dominant retinitis pigmentosa: four new mutations in rhodopsin, one of them in the retinal attachment site. *Genomics* **11**, 199–205.

39. Sung, C-H., Davenport, C.M., Hennessey, J.C., Maumenee, I.H., Jacobson, S.G., Heckenlively, J.R., Nowakowski, R., Fishman, G. Gouras, P. & Nathans, J. (1991) Rhodopsin mutations in autosomal dominant retinitis pigmentosa. *Proc. Natl. Acad. Sci. U.S.A.* **88**, 6481–6485

40. Inglehearn, C.F., Keen, T.J., Bashir, R., Jay, M., Fitzke, F., Bird, A.C., Crombie, A. & Bhattacharya, S.S. (1992) A completed screen for mutations of the rhodopsin gene in a panel of patients with autosomal dominant retinitis pigmentosa. *Hum. Mol. Genet.* **1**, 41–45

41. Farrar, G.J., Findlay, J.B.C., Kumar-Singh, R., Kenna, P., Humphries, M.M., Sharpe, E. & Humphries, P. (1992) Autosomal dominant retinitis pigmentosa: a novel mutation in the rhodopsin gene in the original 3q linked family. *Hum. Mol. Genet.* **1**, 769–771

42. Dryja, T.P., Berson, E.L., Rao, V.R. & Oprian, D.D. (1993) Heterozygous missense mutation in the rhodopsin gene as a cause of congenital stationary night blindness. *Nat. Genet.* **4**, 280–283

43. Davies, K. (1993) Peripherin and the vision thing. *Nature (London)* **362**, 92

44. Nathans, J., Davenport, C.M., Maumenee, I.H., Lewis, R.A., Hejtmancik, J.F., Litt, M., Lovrien, E., Weleber, R., Bachynski, B., Zwas, F., *et al.* (1989) Molecular genetics of human blue cone monochromacy. *Science* **245**, 831–838

45. Nathans, J. (1987). Molecular biology of visual pigments. *Annu. Rev. Neurosci.* **10**, 163–194

46. Nathans, J. (1990) Determinants of visual pigment absorbance: role of charged amino acids in the putative transmembrane segments. *Biochemistry* **29**, 937–942

47. Honig, B., Dinur, U., Nakanishi, K., Balogh-Nair, V., Gawinowicz, M.A., Arnaboldi, M. & Motto, M.G. (1979) An external point-charge model for wavelength regulation in visual pigments. *J. Am. Chem. Soc.* **101**, 7084–7086

48. Neitz, M., Neitz, J. & Jacobs, G.H. (1991) Spectral tuning of pigments underlying red-green color vision. *Science* **252**, 971–974

49. Winderickx, J., Lindsey, D.T., Sanocki, E., Teller, D.Y., Motulsky, A.G. & Deeb, S.S. (1992). Polymorphism in red photopigment underlies variation in colour matching. *Nature (London)* **356**, 431–433

50. Merbs, S.L. & Nathans, J. (1992) Absorption spectra of human cone pigments. *Nature (London)* **356**, 433–435

51. Merbs, S.L. & Nathans, J. (1992) Absorption spectra of the hybrid pigments responsible for anomalous color vision. *Science* **258**, 464–466

52. Mollon, J.D. (1992) Mixing genes and mixing colours. *Curr. Biol.* **3**, 82–85

53. Oprian, D.D., Asenyo, A.B., Lee, N. & Pelletier, S.L. (1991) Design, chemical synthesis and expression of genes for the three human color vision pigments. *Biochemistry* **30**, 11367–11372

54. Okano, T., Fukada, Y., Artamonov, I.D. & Yashikawa, T. (1989) Purification of cone visual pigments from chicken retina. *Biochemistry* **28**, 8848–8856

55. Kuwata, O., Imamoto, Y., Okano, T., Kokame, K., Kojima, D., Matsumoto, H., Morodome, A., Fukada, Y., Shichida, Y., Yasuda, K., Shimura, Y. & Yoshizawa, T. (1990) The primary structure of iodopsin, a chicken red-sensitive cone pigment. *FEBS Lett.* **272**, 128–132

56. Tokunaga, F., Iwasa, T., Miyagishi, M. & Kayada, S. (1990) Cloning of cDNA and amino acid sequence of one of chicken cone visual pigments. *Biochem. Biophys. Res. Commun.* **173**, 1212–1217

57. Ovchinnikov, Y.A. Abdulaev, N.G., Zolotarev, A.S., Artamonov, I.D., Bespalov, I.A., Dergachev, A.E. & Tsuda, M. (1988) Octopus rhodopsin. Amino acid sequence deduced from cDNA. *FEBS Lett.* **232**, 69–72

58. Robinson, J., Schmitt, E.A., Harosi, F.I., Reece, R.J. & Dowling, J.E. (1993) Zebrafish ultraviolet visual perception: absorption spectrum, sequence and localization. *Proc. Natl. Acad. Sci. U.S.A.* **90**, 6009–6012

6

The roles of molecular chaperones *in vivo*

Peter A. Lund

School of Biological Sciences, University of Birmingham, Birmingham B15 2TT, U.K.

Introduction

There has been a recent explosion of interest in the question 'How do proteins fold in the cell?' This is because current models of protein folding *in vitro* are rather inadequate when extrapolated to the cytosol. Among the factors that differ between the cosy world of the test-tube and the chaos of the cell are the concentrations at which folding reactions occur; the fact that one end of a protein is synthesized and folded before the other; the need to keep some proteins unfolded or partially folded for them eventually to attain their biological function and position; and the fact that the cell may have to cope with conditions which are far from optimal for protein folding, such as heat stress[1]. At the forefront of research in this area are the molecular chaperones, which can be defined as proteins which assist other proteins to reach their final active form. The mechanisms by which they do this have been the subject of much study (see Chapter 7, p. 125). It is important when considering mechanisms of molecular chaperones to bear their properties *in vivo* in mind, and also to remember the diversity of their roles. The aim of this review is to present some of the evidence for the cellular function of the molecular chaperones. Two families of heat shock proteins (hsp) — hsp70 and hsp60 — will be discussed in some detail, and other examples of chaperones that interact non-covalently with their targets will also be mentioned. Whether or not those proteins which covalently modify other proteins and thus help them to fold should be considered as chaperones is a matter of definition, but they will not be included here.

The need for molecular chaperones was not recognized for many years, principally because of the success of experiments which demonstrated the complete unfolding and refolding of proteins *in vitro*[2]. The moral of this tale — that information from purified systems may only tell part of the story of the more complex cellular milieu — can still be applied in this field. The demonstration *in vitro* that a protein can function as a molecular chaperone does not prove that it does so *in vivo*. While it is relatively easy to demonstrate such a property for a purified protein, it is harder to show that this is indeed the role of the protein in the cell. Moreover, molecular chaperones which share the common biochemical property of chaperoning protein folding may, nevertheless, carry out different functions within the cell. Evidence for the function of a chaperone *in vivo* has to come principally from studies where it is expressed in an altered form, or is missing altogether — in other words, from genetic studies. A consequence of depletion of the chaperone should be that some proteins fail to reach their active state. In addition, experiments should probe the interactions of putative chaperones with their substrates *in vivo*: direct evidence for association with partly folded proteins is desirable. Both of these approaches can be problematical: the former because many chaperones are apparently essential for cell function, the latter because such interactions are transient. With these caveats in mind, however, a good deal can be deduced from the available data.

The hsp70 proteins: multi-functional and ubiquitous chaperones

Initially discovered as proteins whose synthesis was increased on heat-shock, the hsp70 proteins are now recognized as a large, highly conserved and ubiquitous group, not all of which are heat inducible. *Escherichia coli* contains a heat-inducible hsp70 called DnaK. *DnaK* genes in other bacteria have not been extensively studied, but most bacteria contain only one. In the yeast *Saccharomyces cerevisiae*, however, the situation is far more complex; there are at least six hsp70 homologues in the cytosol (SSA1–SSA4, SSB1 and SSB2), and in the mitochondria (SSC1p) and in the endoplasmic reticulum (Kar2)[3]. Studies in other eukaryotic cells show that hsp70 proteins can migrate to the nucleus; they are present in chloroplasts, and their presence in other organelles has been postulated.

Genetic studies show that these proteins are important and sometimes essential for cell function. Although *E. coli* can survive the loss of its *dnaK* gene, such cells are distinctly 'distressed': they only grow at low temperatures and tend to form filaments. In yeast, at least one of the *ssa* genes must be functional and expressed at high levels for viability, and although an *ssb1–ssb2* double mutant can be made, it grows slowly. The SSA proteins appear to have a role in protein translocation across membranes, since depletion of SSA protein from yeast cells leads to an accumulation of precursors of both mitochondrial and secreted proteins[3]. In this context it is noteworthy that, in *E. coli*, over-expression of DnaK has been observed to stimulate the membrane translocation of a hybrid protein which is usually unable to cross the inner membrane[4]. The *ssb* gene products are found to be strongly associated with ribosomes, and may be associated with newly translated proteins. Similar associations between nascent protein and hsp70 have been observed in mammalian cell lines by immunoprecipitation experiments[5], and a role of hsp70 in binding to proteins as they emerge from the ribosome — perhaps to prevent their aggregation or misfolding, or to act as a precursor for subsequent transfer to the hsp60 proteins — seems very likely.

The roles of the mitochondrial and endoplasmic reticulum proteins are also likely to be consequences of the interaction of hsp70 with partially folded proteins. Many studies have shown that proteins need to be in an unfolded state to be able to cross membranes[6], and it seems likely that hsp70 homologues on both sides of the membrane maintain protein in this state for membrane transport to occur. The yeast mitochondrial protein SSC1p is essential for cell viability[3], so studies have been limited to the use of temperature-sensitive alleles; when yeast strains containing these are shifted to the non-permissive temperature, several mitochondrial proteins which are normally imported from the cytosol accumulate as precursors, probably trapped in the mitochondrial membrane. The *kar2* gene [coding for the yeast homologue of the well-known mammalian immunoglobulin heavy-chain-binding protein (BiP)] is also essential for growth, and studies on temperature-sensitive alleles of the gene have shown similar accumulation of protein precursors[3]. Mammalian BiP also has properties that point to a molecular chaperone role: it is induced by the presence of unfolded proteins in the lumen of the endoplasmic reticulum, and complexes between nascent secretory proteins and BiP can be isolated from cells[1].

What is the role of hsp70 in stressed cells? The ability of hsp70 to bind to unfolded proteins *in vitro* has led to the suggestion that one of their roles *in vivo* is the protection of partially denatured proteins[7], although they may also target some denatured proteins to the degradative pathways. More strikingly, it has recently been demonstrated that they play a role in reactivation of damaged proteins. This was shown with the enzyme luciferase, which is rapidly inactivated in *E. coli* that is heated to 43 °C but recovers activity on return of the cells to 37 °C[8]. However, this reactivation is substantially reduced in *dnaK* mutant cells. A similar reduction is also seen with mutants of the *dnaJ*, *groES* and *groEL* genes, as discussed below.

The original genetic screen which identified the *dnaK* gene in *E. coli* (inability to plate bacteriophage λ) also revealed a second gene called *dnaJ* which was shown to be part of the same operon. An interaction between DnaK and DnaJ proteins was proposed on the grounds that some *dnaK* mutants could be suppressed by mutations in *dnaJ*; such an interaction has been confirmed biochemically. Proteins with motifs typical of the DnaJ protein are found in all organisms. In yeast, there are at least four genes which code for proteins with DnaJ motifs, and defects in one of these (*ydj1*) leads to a phenotype similar to that resulting from defects in the *ssa* genes, suggesting that interactions between hsp70 and DnaJ homologues are required for the membrane translocation of proteins. One protein with a DnaJ motif, Sec63, was originally identified in a screen for secretion-defective mutants of yeast, and is known to be an endoplasmic reticulum membrane protein which interacts with the product of the *kar2* gene. As predicted from the lumenal location of Kar2, the DnaJ motif in this membrane-bound protein faces the lumen of endoplasmic reticulum[3].

Thus evidence from genetic and other studies *in vivo* show that the hsp70 proteins are involved in protein translocation across membranes, protein translation, and protection and recovery from heat shock. Other roles, such as uncoating of clathrin cages, initiation of DNA replication and sensing the temperature of the cell, have also been reported or suggested. Interaction with proteins with DnaJ motifs appears to be an important part of at least some of these processes. All these seem likely to stem from the interaction of hsp70 with unfolded or partially folded precursors and the ability of DnaJ to modulate these interactions. A precise explanation of their role in terms of their molecular mode of action is still not available.

The hsp60 proteins: archetypal molecular chaperones

The hsp60 proteins were the first proteins to be recognized as having a chaperone function[9]. As with the hsp70 proteins, they are encoded by highly conserved genes and are found in all organisms and many cellular locations. The similarity between organellar and cytosolic members of this family is lower than between the hsp70 homologues, and they will be treated independently here.

The GroE-type proteins

The genetic screen for mutants in *E. coli* that failed to plate bacteriophage λ led to the identification of two proteins, GroEL and GroES, that were essential for this process. The genes for these proteins are transcribed as a single mRNA which is present in all cells but is highly induced on heat shock[10]. The proteins are known to assemble into large complexes *in vivo*. Unlike *dnaK*, the *groE* genes are essential for cell viability. A striking demonstration of their role in normal cellular growth was obtained by experiments in which expression of all the heat-shock genes (about 20 in *E. coli*) was prevented or substantially lowered by deletion of the subunit of RNA polymerase that is responsible for transcription of these genes. Such cells grow poorly, and only at low temperatures. However, mutants of these strains could be found which grew at close to normal temperatures: these were due to mutations which greatly elevated the constitutive levels of expression of the *groESL* genes[11]. Thus, even at normal growth temperatures, the GroES and GroEL proteins play crucial roles.

Experiments on temperature-sensitive alleles of these genes revealed a wide range of pleiotropic effects on protein synthesis and degradation, DNA and RNA replication, recombination, and other cellular processes[10]. As mentioned above, *groEL* and *groES* mutants are also defective in the reactivation of a heat-inactivated protein *in vivo*[8]. The problem with interpretation of these results arises from the fact that any protein which is involved in protein folding may affect any process in the cell that require active enzymes, so these data merely support the assertion that the GroE proteins play an essential role. Evidence that this role is indeed concerned with protein folding came initially from the demonstration that a protein whose active expression in *E. coli* was normally poor could be expressed at high levels if the GroES and GroEL proteins were simultaneously over-produced[12]. It was this demonstration that led directly to the many biochemical studies on this protein reviewed by Burston and Clarke (see Chapter 7, p.125). Other evidence also showed that the GroE proteins interacted with several proteins *in vivo*; in particular the fact that over-expression of GroES and GroEL could suppress the temperature-sensitive phenotype of several distinct and structurally unrelated mutant proteins[13].

Presumably, the mutants failed to fold into active protein at the non-permissive temperature because some step in their folding pathway had become temperature sensitive, and the higher levels of GroES and GroEL protein were able to overcome this defect. The most recent evidence of an interaction between the GroE proteins and active protein precursors comes from experiments where the normal *groEL* gene has been mutated to be highly temperature sensitive, losing its function at 37 °C. Cells expressing this mutant protein, which have been grown at 30 °C and then shifted to 37 °C, eventually cease to grow; the profile of soluble protein synthesized after this temperature shift is strikingly different to wild-type cells, with many identifiable proteins disappearing[14]. This is the most convincing demonstration to date that endogenous proteins do associate with GroEL *in vivo*; however, a direct demonstration of this is still lacking. It is to be hoped that this approach will eventually answer a question which has long been prevalent in this field: do all newly synthesized proteins bind to GroEL or is the chaperone pathway for some proteins only? If this can be answered, then some start can be made to answering the question of what are the features of those proteins which do bind chaperones that enable them to do so?

The position of GroEL and GroES in the pathway of protein folding in *E. coli in vivo* and *in vitro* is still not clear, although evidence *in vitro* strongly suggests that they can accept proteins which have been previously bound by DnaK and DnaJ[15]. (That this is not obligatory for protein folding is shown by the viability of *dnaK* mutant cells.) Studies on mitochondria also support the assertion that hsp70 and hsp60 act as part of a protein-folding pathway, with proteins being transferred from hsp70 to hsp60. Hsp60 proteins are found in mitochondria and chloroplasts. They show a high degree of sequence similarity with bacterial GroEL proteins, consistent with the endosymbiotic hypothesis of mitochondrial origin. Studies on *S. cerevisiae* and *Tetrahymena* have shown that the hsp60 proteins are essential for cell growth. As before, the use of temperature-sensitive alleles in the genes has been informative; at the non-permissive temperature mitochondria accumulate inactive and unfolded forms of several imported proteins[3]. However, unlike the situation in the hsp70-defective mitochondria (where precursors remained trapped in transit), the proteins are fully imported into the mitochondria[3]. Thus hsp60 in mitochondria must act later than hsp70.

The cytosolic TCP1-like proteins

The essential and central nature of the GroEL and GroES proteins in protein synthesis in bacteria prompted searches for homologues in the eukaryotic cytosol, that being the principle site of protein synthesis. A class of proteins that showed weak but significant homology to the GroEL protein was observed, typified originally by a mouse protein called TCP1. These proteins are found in all cells, and homologues have now been cloned from several organisms, including yeast[3]. Homologues appear to assemble into a hetero-oligomeric protein complex, called TRiC (TCP1 ring complex), the precise constituents of which remain to be fully described. Several lines of biochemical evidence suggest that this complex is a cytosolic chaperone, but studies *in vivo* have so far been quite limited. A gene for a component of the protein has been identified in yeast, and shown to be essential[3,16]. Studies on strains carrying conditional mutants of this gene have shown that they fail to assemble microtubules. However, the role of the TRiC complex is still not clearly defined by studies *in vivo*, and this is a very active area of current research. Questions that apply to the GroE proteins apply even more so here: do all proteins have to be chaperoned by TRiC, or only a subset? Does TRiC have any functions other than chaperoning the folding of newly folded proteins; for example, is it involved at all in membrane translocation? Is there a counterpart in the endoplasmic reticulum which is still awaiting discovery? Why does TCP1 appear not to be heat-shock inducible? Do other proteins fulfil the central role that GroEL and GroES play in bacteria in this respect?

Other molecular chaperones: probable and possible

The hsp90 proteins

Another abundant and ubiquitous class of heat-shock proteins which appear to have some properties typical of molecular chaperones *in vivo* are in the 80–90 kDa range, and are generally referred to as the hsp90 proteins. Studies on the *E. coli* homologue (actually a diminutive member of this family at 62.5 kDa) have been rather uninformative; the gene helps cells to grow more efficiently at high temperature, but is not essential for growth. However, in *S. cerevisiae*, which has two genes for this family, the transcription of at least one is essential for growth. Temperature-sensitive alleles have not yet been described and so the effect of inactivation of the hsp90 proteins is unknown. Nevertheless, a variety of studies on various cell lines have shown that the hsp90 proteins have the property of binding many hormone receptors and other important cellular proteins in an inactive form, and releasing them in an active form when the appropriate ligand is present[17]. This is reminiscent of a molecular chaperone type action: the binding of unfolded and inactive protein, followed by its release in an altered form. Studies *in vitro* have shown that hsp90 does indeed display molecular chaperone properties; the precise relationship of these properties to its role *in vivo* remains to be established.

The rest: a large and growing group

A literature search using the key-word 'chaperone' produces a lot of references. There is a tendency for proteins to be labelled 'molecular chaperones' rather indiscriminately, when clear evidence for their function *in vivo* and mechanism *in vitro* is still lacking. Nevertheless, it is becoming clear that many proteins may possess the correct entry requirements for membership of the chaperone club. For example, the SecB protein of *E. coli* has long been known to bind to precursors of certain secreted proteins, maintaining them in an unfolded form and guiding them to the membrane-translocation apparatus. Depletion of SecB leads to the accumulation of the cytosolic precursors of these proteins. SecB is highly specific in its interactions with other proteins[18] and thus illustrates the fact that some chaperones are more selective than others when it comes to substrates. The extreme example of this is shown by chaperones that are specific to only one or two substrate proteins: these are best typified by the periplasmic bacterial protein PapD. *PapD* mutants fail to assemble the long, polymerized bacterial pili, the subunits of which are instead rapidly degraded[19]. Complexes between PapD and its substrates — such as the adhesin, PapG, which is situated at the tip of the pilus — can be isolated from growing cells. It does seem to be the case, however, that the proteins with which PapD associates are more fully folded than the substrates of the more promiscuous chaperones, such as those discussed above. Therefore, the mechanism of action of PapD may be very different from that of hsp60 and hsp70. Many other examples of proteins involved in protein secretion, assembly, and repair which may well be chaperones could be cited; space precludes a fuller treatment of these.

Summary

Table 1 summarizes the families of chaperones mentioned in this review, and lists their proposed functions. Many of these proteins are named in the accompanying review of Burston and Clarke.

Table 1. A summary of some of the proteins for which there is good evidence of molecular chaperone function *in vivo*

Name of chaperone or chaperone family and examples	Proposed function(s)
hsp70 family	
DnaK (bacteria)	Essential for growth at high and low temperatures and required for healthy growth at all temperatures
	Required for induction of heat-shock response
	Reactivates heat-damaged proteins
	Involved in cell division
	Role in secretion of some periplasmic proteins
	Acts with DnaJ/GrpE
	Essential for λ DNA replication initiation
Cytosolic hsp70s: SSA1–SSA4; SSB1, SSB2 (yeast); hsp70 and hsc70 (animal cells)	Import of proteins into mitochondria, chloroplasts and endoplasmic reticulum
	Disassembly of clathrin cages
	Binding of nascent protein chains
	Protection against and recovery from heat shock
Endoplasmic reticulum hsp70s: Kar2 (yeast); BiP (mammals)	Binding of immunoglobulin light chains and other multisubunit proteins in endoplasmic reticulum
Mitochondrial and plastid hsp70s	Binding and folding of imported protein precursors
DnaJ-type proteins	Interact with hsp70 proteins to prevent aggregation of unfolded or nascent proteins
	May have a role in secretion and in reactivation of heat-inactivated proteins

Name of chaperone or chaperone family and examples	Proposed function(s)
hsp60 family	
GroEL-like proteins	Binding of unfolded or partially folded proteins; prevention of protein aggregation
(GroEL in bacteria, hsp60 in mitochondria and chloroplasts)	May be involved in reactivation of heat-damaged proteins
	Required for folding of newly imported mitochondrial and chloroplast proteins
	Act with co-chaperones: GroES in bacteria; homologous proteins found in mitochondria and chloroplasts
TCP1-like proteins	Folding of actin and tubulin and possibly other cytosolic proteins
hsp90	Binds hormone receptors and other proteins in inactive form; released when ligand present
SecB	Required for secretion of a subset of bacterial proteins: binds them in unfolded form and delivers them to membrane translocation machinery
PapD and related proteins	Periplasmic family of proteins which act in the assembly of subunits of specific bacterial pili or fimbriae
Calnexin (IP90)	Trans-endoplasmic reticulum-membrane protein which associates transiently with several different proteins including class II MHC and T-cell receptor; probably required for their functional assembly into membrane complexes

This list is not exhaustive. Abbreviations used: hsc, constitutively-synthesized protein with high homology to an hsp; MHC, major histocompatibility complex.

- *Molecular chaperones are proteins which interact with other proteins and help them to reach their final, active conformation. They appear to do this by binding them in an unfolded or partially folded state and subsequently releasing them in an altered form. This property may endow them with several essential or important roles in addition to helping newly synthesized proteins to fold correctly, such as repairing damaged proteins and assisting proteins in membrane translocation. To confirm that a given protein has molecular chaperone activity in vivo, it is necessary to show that interactions between the chaperone and other proteins do occur in the cell, and that loss of the molecular chaperone leads to the accumulation of inactive or precursor protein.*

- *The hsp70 protein family are highly conserved and ubiquitous. Genetic studies confirm that their depletion leads to the accumulation of inactive precursor or other proteins, and immunochemical studies show they associate with nascent polypeptides. They are implicated not only in protein folding, but also in protein transport across membranes and reactivation of heat-damaged proteins.*

- *The hsp60 proteins are also ubiquitous and very similar in sequence. Those found in bacteria and organelles, such as mitochondria (the GroEL family), are essential at all temperatures, and particularly after heat shock. Their loss or depletion leads to the formation of protein aggregates and eventual cell death. A co-chaperone protein (GroES) is required for their function. Cytosolic homologues (the TCP1 family) are also essential, though not heat-shock induced; they are believed to have a chaperone role in tubulin assembly and their actual role in the cell may be much broader.*

- *Many other proteins may have a chaperone function in vivo. Such a function may be specific to a particular substrate (such as the PapD protein in E. coli); others may be more general (such as hsp90 and SecB). Evidence is still needed to demonstrate whether all those proteins which show chaperone behaviour in vitro actually have such a role in vivo. It seems likely that different classes of chaperone may overlap in their specificity, and it is certain that the various proteins classed as molecular chaperones fulfil a wide variety of roles in the cell.*

References

1.   Gething, M.-J. & Sambrook, J. (1992) Protein folding in the cell. *Nature (London)* **355**, 33–45
2.   Anfinsen, C.B. (1973) Principles that govern the folding of polypeptide chains. *Science* **181**, 223–230
3.   Craig, E.A., Gambill, B.D. & Nelson, R.J. (1993) Heat shock proteins: molecular chaperones of protein biogenesis. *Microbiol. Rev.* **57**, 402–414
4.   Phillips, G.J. & Silhavy, T.J. (1990) Heat-shock proteins DnaK and GroEL facilitate export of Lac Z hybrid proteins in *E. coli. Nature (London)* **344**, 882–884

5.   Beckmann, R.P., Mizzen, L. & Welch, W. (1990) Interaction of hsp70 with newly synthesised proteins: implications for protein folding and assembly. *Science* **248**, 850–856

6.   Eilers, M. & Schatz, G. (1988) Protein unfolding and the energetics of protein translocation across membranes. *Cell* **52**, 481–483

7.   Pelham, H.R.B. (1986) Speculations on the functions of the major heat shock and glucose regulated proteins. *Cell* **46**, 959–961

8.   Schroeder, H., Langer, T., Hartl, F.-U. & Bukau, B. (1993) DnaK, DnaJ and GrpE form a cellular chaperone machinery capable of repairing heat induced damage. *EMBO J.* **12**, 4137–4144

9.   Hemmingsen, S.M., Woolford, C., van der Vies, S.M., Tilly, K., Dennis, D., Georgopoulos, C., Hendrix, R.W. & Ellis, R.J. (1988) Homologous plant and bacterial proteins chaperone oligomeric protein assembly. *Nature (London)* **333**, 330–334

10. Zeilstra-Ryalls, J., Fayet, O. & Georgopoulos, C. (1991) The universally conserved GroE (hsp60) chaperonins. *Annu. Rev. Microbiol.* **45**, 301–325

11. Kusukawa, N. & Yura, T. (1988) Heat shock protein GroE of *Escherichia coli:* key protective roles against thermal stress. *Genes Dev.* **2**, 874–882

12. Goloubinoff, P., Gatenby, A.A. & Lorimer, G.H. (1989) GroE heat-shock proteins promote assembly of foreign prokaryotic ribulose bisphosphate carboxylase oligomers in *Escherichia coli.* *Nature (London)* **337**, 44–47

13. Van Dyk, T.K., Gatenby, A.A. & LaRossa, R.A. (1989) Demonstration by genetic suppression of interaction of groE products with many proteins. *Nature (London)* **342**, 451–453

14. Gragerov, A., Nudler, E., Komissarova, N., Gaitnaris, G., Gottesman, M.E. & Nikiforov, V. (1992) Cooperation of GroEL/GroES and DnaK/DnaJ heat shock proteins in preventing protein misfolding in *Escherichia coli.* *Proc. Natl. Acad. Sci. U.S.A.* **89**, 10341–10344

15. Horwich, A.L., Low, K.B., Fenton, W.A., Hirshfield, I.N. & Furtak, K. (1993) Folding *in vivo* of bacterial proteins: role of GroEL. *Cell* **74**, 909–917

16. Hendrick, J.P. & Hartl, F.-U. (1993) Molecular chaperone functions of heat-shock proteins. *Annu. Rev. Biochem.* **62**, 349–384

17. Pratt, W.B. (1990) Interaction of hsp90 with steroid receptors: organizing some diverse observations and presenting the newest concepts. *Mol. Cell Endocrinol.* **74**, 69–76

18. Kumamoto, C.A. & Francetic, O. (1992) Highly selective binding of nascent polypeptide by an *Escherichia coli* chaperone protein in vivo. *J. Bacteriol.* **175**, 2184–2188

19. Hultgren, S.J., Normark, S. & Abraham, S.N. (1991) Chaperone-assisted assembly and molecular architecture of adhesive pili. *Annu. Rev. Microbiol.* **45**, 383–415

7

Molecular chaperones: physical and mechanistic properties

Steven G. Burston and Anthony R. Clarke

Department of Biochemistry and Molecular Recognition Centre, University of Bristol, School of Medical Sciences, University Walk, Bristol BS8 1TD, U.K.

Introduction

Since the early 1960s protein folding has been the focus of intense experimental and theoretical interest, initiated largely by the observation that bovine ribonuclease can refold spontaneously from a structureless and inactive state. This led Anfinsen[1] to propose that all of the information necessary for a protein to fold into its correct 3-dimensional structure is contained within its amino acid sequence. This hypothesis is strongly supported by the observation that many denatured proteins fold spontaneously and efficiently *in vitro* when placed in renaturing conditions but leaves us with the puzzle that many others do not, often forming insoluble aggregates[2]. Although refolding under non-physiological conditions, such as at low temperatures or at very low protein concentrations, can sometimes improve yields of native material, it is difficult to see how this is achieved in the cell. However, since it is essential *in vivo* that proteins are both transported to their correct sites and are able to fold and assemble when they arrive, it is reasonable to expect that the cell has acquired a means to achieve this with high efficiency. During the last decade this expectation has been realized with the discovery of a new class of *in vivo* 'folding helpers', known collectively as molecular chaperones (see Chapter 6, p. 113). These are defined as 'a family of unrelated classes of protein that mediate the correct folding and assembly of other polypeptides, but are not themselves components of the final functional structure'[3]. The bulk of the molecular chaperones also belong to the set of heat-shock proteins (hsps), so-called because of their elevated expression under conditions of cellular stress. In this review we describe the mechanistic information gained from studies *in vitro* of the hsp70, hsp60, TCP1 and hsp90 proteins. The molecular chaperones also include two families of proteins which catalyse chemically defined processes in protein folding, namely peptidylprolyl isomerase and protein disulphide-isomerase. The reader should consult reviews by Schmid and Freedman for a detailed description of the functions of these enzymes *in vivo* and *in vitro*[2].

The hsp70 molecular chaperone

Hsp70 homologues can be found both in bacteria and in the cytoplasm, mitochondria, endoplasmic reticulum (ER) and chloroplasts of eukaryotic cells where they are expressed both constitutively and at higher levels under stress. Studies of hsp70 action *in vivo* have shown that, among its many cellular functions, it binds to nascent polypeptide chains emerging from the ribosome and maintains polypeptide chains targeted to organelles in an unfolded, translocation-competent state[4]. Under heat-shock conditions, hsp70 was found to accumulate largely in complexes with unfolded, nascent ribosomal proteins. This affinity for unfolded proteins can be reversed by the addition of Mg^{2+} and ATP. It was originally suggested from this initial work *in vivo* that hsp70 functioned by binding to unfolded proteins via exposed, hydrophobic amino acid side-chains, and that the ATP-dependent release of bound proteins allowed them to fold[5]. This simple function of hsp70 has a large number of uses within the cell where unfolded polypeptide chains need to be stabilized to prevent misfolding and aggregation and to allow translocation across membranes.

The hsp70 consists of two domains: the highly conserved *N*-terminal domain is a weak ATPase, while the less-well-conserved *C*-terminal domain is responsible for the binding of the substrate protein. A 44 kDa chymotryptic fragment containing the ATPase domain has been crystallized and the X-ray structure determined to a resolution of 2.2 Å[6]. This ATPase fragment has a fold identical to the globular G-actin monomer and to hexokinase, both of which undergo a conformational change during their reaction cycles, thus suggesting ATP binding and hydrolysis may cause similar conformational changes in hsp70. This has been confirmed by changes in the pattern of tryptic fragments produced by ATP binding. Observations on the conditions needed to displace bound protein substrates suggest that there are two conformational states of hsp70: the ATP-bound state, which has a low affinity for unfolded protein substrates, and an ADP-bound state, which has a much higher affinity (see Figure 1). The kinetics of ATP hydrolysis showed that the binding of unfolded protein stimulates the ATPase activity and that the dissociation of products is slow compared with the hydrolytic step, so the predominant complex in the catalytic cycle is hsp70–ADP, i.e. the tight-binding form[7]. The

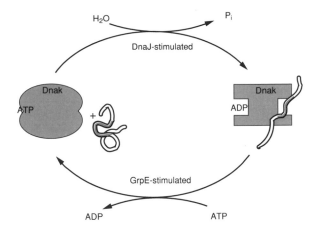

Figure 1. The reaction cycle of hsp70 and its co-proteins
The scheme shows the interconversion of two conformers of hsp70. The ATP state (left) binds proteins weakly but, on conversion to the ADP state, it interacts strongly with hydrophobic regions (blue) of unfolded proteins. Hydrolysis is stimulated by DnaJ and exchange of ADP for ATP is promoted by the binding of unfolded protein chains and GrpE.

increased rate of the ATP hydrolytic cycle in the presence of substrate protein is attributed to stimulation of ADP dissociation, thus allowing the rebinding of ATP. More recent work has examined the dissociation of complexes between DnaK (the *Escherichia coli* hsp70 homologue) and protein substrates using high-performance liquid chromatography (h.p.l.c.) gel-exclusion chromatography in the presence of different nucleotides and cations[8]. It was concluded that dissociation of the protein substrate from DnaK requires K^+, that Mg–ATP binding and not hydrolysis is necessary for substrate dissociation, and that release of the substrate protein from the complex must be rapid compared with the rate of ATP hydrolysis. This switching between two conformational states with different affinities allows controlled binding and release of substrate proteins.

A structure for the C-terminal domain has been proposed, based upon homology modelling onto the X-ray structure of the soluble domain of HLA-A2 (human leukocyte associated antigen) of human major histocompatibility complex (MHC). In this predicted structure, the peptide-binding groove of the MHC is lined with hydrophobic residues when the hsp70 amino acid sequence is modelled into the MHC fold, supporting the view that hsp70 binds its substrate via hydrophobic interactions. This has been demonstrated further by the finding that, when challenged with a set of random heptapeptides, BiP (human immunoglobulin heavy-chain-binding protein, the form of hsp70 in the ER) showed a preference for a hydrophobic residue at every position in the peptide sequence, and that such sequences would statistically occur regularly (every 16 residues) in a globular protein[9]. The structure of bound substrates has also been examined using biophysical techniques. Reduced, carboxy-methylated α-lactalbumin, which has little or no secondary structure as judged using far-u.v. circular dichroism, binds tightly to both DnaK and mammalian hsp70. This is strongly supported by nuclear magnetic resonance (n.m.r.) analysis of peptides binding to DnaK which indicates that they bind in a fully extended structure.

Hsp70 as part of a larger molecular chaperone complex

There is strong evidence to suggest that hsp70 may interact with other protein co-factors for its function *in vivo*. In *E. coli*, DnaJ (a dimer of 41 kDa subunits) and DnaK are encoded in the same operon and their interaction is suggested by mutational analysis. This interaction has been demonstrated *in vitro*, where it was found that DnaJ stimulated the hydrolytic step of DnaK[7]. Since the rate of dissociation of ADP is the slowest step, addition of DnaJ will promote the formation of the DnaK–ADP complex and thus increase the proportion of molecules in the state with a high affinity for the substrate protein. This tight protein-binding state has been demonstrated using unfolded rhodanese, a mitochondrial protein that refolds poorly, as a substrate[10]. A further hsp encoded by the *E. coli* gene *grpE* has been shown to be a part of this chaperone complex. GrpE (a 22 kDa monomer) stimulates the DnaK ATPase 50-fold in the presence of DnaJ by enhancing the rate of ADP–ATP exchange on DnaK, presumably by triggering the release of ADP[7]. The effect of increasing this nucleotide exchange rate is to promote the formation of the ATP state, thus lowering the affinity of DnaK for protein substrate and allowing its dissociation. This mechanism (see Figure 1) is supported by the observation that GrpE causes rhodanese to dissociate from the DnaK–DnaJ complex. It has been suggested that unfolded proteins are then transferred to an hsp60 (such as the *E. coli* chaperonins) to facilitate folding to the native state[4].

DnaK, DnaJ and GrpE have also been shown to increase the folding efficiency of a protein *in vitro*. Firefly luciferase can be thermally inactivated at 42 °C whereupon the protein forms insoluble aggregates. In the presence of DnaK and DnaJ, the enzyme unfolds and forms a stable ternary complex with DnaK–DnaJ, even in the presence of ATP and GrpE. When the temperature is reduced to 30 °C, luciferase activity returns. This regaining of activity is dependent upon ATP and all of the three molecular chaperones[4].

The role of hsp70 in maintaining a translocation-competent protein conformation

The involvement of hsp70 in the import of proteins into mitochondria has been demonstrated by analysis of a temperature-sensitive mutant of the yeast mitochondrial hsp70, SSC1p, which results in the accumulation of incompletely translocated polypeptides. In a translocation system *in vitro*, addition of a completely unfolded protein resulted in more efficient translocation than addition of partially or completely folded protein. Therefore, targeted proteins need to be maintained in an extended conformation by cytosolic hsp70 to allow efficient translocation. Inside the mitochondrion, the organellar hsp70

binds to the incoming polypeptide chain to provide the energy for translocation[11]. It is likely that this vectorial driving force is provided by the greater
binding affinity of the mitochondrial hsp70 than the cytosolic equivalent,
possibly due to the fact that the [ADP]/[ATP] ratio is higher inside the
mitochondria than within the cytosol. The imported protein is then folded and
assembled in a process involving the mitochondrial hsp60.

Translocation into the ER also provides a role for BiP, the mammalian ER
hsp70 homologue[4]. Temperature-sensitive mutants of the yeast homologue of
BiP, Kar2p, also accumulate partially translocated polypeptide chains. The driving force for translocation to the ER in yeast may be interaction between the
imported polypeptide chain and a complex involving BiP and the membrane-
associated Sec63p, which has homology to DnaJ. The role of BiP in subsequent
folding and assembly is probably similar to the role played by hsp70 in the
cytosol, although ER homologues of GrpE and hsp60 have yet to be found.

The action of chaperonins in assisted protein folding

The chaperonins, or hsp60 proteins, are the most intensively studied of the
molecular chaperones. This popularity can be accounted for by their abundance and ubiquitous distribution, their intriguing architecture and the fact
that they were the first identified factors which facilitated protein folding[3].
They are found in the cytoplasm of both prokaryotes and eukaryotes, in plas-
tids and in mitochondria. The chaperonin found in the eukaryotic cytoplasm is
known as TCP1 and is sufficiently dissimilar in its molecular properties to
warrant a separate section within this article.

These proteins are required for the correct folding and assembly of newly
synthesized proteins and for the recovery of the cell after exposure to either
thermal or chemical stress which denatures protein components. The former
function is satisfied by a high level of constitutive expression and the latter by
their increased synthesis during the heat-shock response. Both activities can be
reproduced *in vitro*, the first by unfolding proteins in denaturants and diluting
them into a renaturing buffer containing purified chaperonins. For proteins
which refold with poor efficiency in buffer alone, the yield of native material is
increased[12,13]. Similarly, to mimic the heat-shock response, thermolabile proteins (which are normally irreversibly denatured by elevated temperatures) are
able to regain activity more effectively in the presence of chaperonins when the
temperature is reduced below the denaturing threshold. Both of these activities
require chemical energy in the form of ATP hydrolysis.

Structurally, chaperonins are among the most curious components of the
cell: they work in pairs. Chaperonin-60 (cpn60) is a large oligomer of 14 subunits, each having a mass of 60 kDa, and chaperonin-10 (cpn10) is a co-protein
composed of seven subunits, each with a mass of 10 kDa. The subunits of
cpn60 are arranged in two stacked 'doughnuts', each having seven subunits in
the ring and a central cavity 5–7 nm in diameter[14]. The subunits of cpn10 form

a single seven-membered ring. The individual proteins can be seen by electron microscopy and associate only when the nucleotides ATP or ADP are present. With ADP, the cpn60–cpn10 complex is asymmetric, the co-protein capping one of the cpn60 rings to give a 14:7 subunit stoichiometry and a shape like a bullet. Such complexes are also seen with ATP, but in some conditions symmetrical complexes are formed in which cpn10 is associated with both ends so that the structure is reminiscent of an American football[15].

Cpn60 subunits have a slow Mg^{2+}- and K^+-dependent ATPase activity. The addition of cpn10 to a mixture of cpn60 and ATP under physiological ionic conditions reduces the hydrolytic rate by half, in accordance with its 'one-sided' association. Furthermore, the 14:7 (cpn60–cpn10) complex formed with ADP is extremely stable, with a rate of dissociation of ADP and cpn10 two orders of magnitude slower than the rate of ATP hydrolysis[16]. This observation explains the halving of the hydrolytic rate through the shut-down of hydrolysis on the 'cpn10 side' during ATP turnover. However, it can be demonstrated that the hydrolysis of ATP on the ATP_7–cpn60–ADP_7–cpn10 complex leads to a rapid release of cpn10 and ADP[17] (S.G. Burston & A.R. Clarke, unpublished work). These arcane kinetic experiments together with the structural studies suggest that, through the cycle of ATP binding and hydrolysis, cpn10 undergoes obligatory binding and dissociation, perhaps on alternate sides of cpn60 and perhaps via a symmetrical intermediate[15,17]. The relevance of such cycles to the function of chaperonins becomes clear on considering their ability to bind and displace folding proteins.

Substrate proteins bind to the cpn60 oligomer only when they are unfolded or partially folded. In the absence of nucleotides and co-protein such complexes are very stable[18]. In general, those proteins that rely heavily on chaperonins to achieve high folding yields bind tightest to produce a complex so stable that it will inhibit the folding reaction. There is no obvious conformational or sequence selection in this binding event, proteins with mainly helical secondary structure, or those which contain only β-sheets, are equally well bound and there are no clearly defined signal sequences. Peptide and protein binding experiments suggest that the complexes are maintained through hydrophobic contacts and that the chaperonin accommodates the bound substrate in the central cavity. Some kinetic[18,19] and recent n.m.r. evidence[20,21] suggests that proteins are bound in highly unfolded conformations, in contrast to earlier observations which suggested that productive folding intermediates were preferentially bound[13]. Irrespective of this question, the addition of ATP weakens the interaction and a combination of ATP and cpn10 weakens it still further, so releasing the bound protein to allow folding[16]. The acquisition of native structure is not achieved in a single cycle of ATP hydrolysis and, for those cases where it has been quantified, chaperonins require the turnover of between 100 and 300 molecules of ATP per folded subunit of substrate. This is something like 10–20% of the energetic cost of covalent synthesis of the protein in these cases[13] (N.J. Dunster & A.R. Clarke, unpublished work).

The cofactor requirement for release depends upon the protein substrate. Some proteins (e.g. bacterial L-lactate dehydrogenase and tryptophanase) need only ATP for dissociation, and more importantly an enhancement of folding yield, while others (e.g. rhodanese and Rubisco) additionally require cpn10. These results show that cpn10 is not an absolute or obligatory requirement for chaperonin-enhanced refolding.

As with other mechanisms of energy transduction, a central question is, 'What is the role played by the nucleotide in modulating the properties of chaperonins?' As stated above, ATP stimulates the release of bound proteins and is hydrolysed during the folding reaction. But is it the binding of ATP or its conversion to ADP and P_i which loosens the interaction with substrates? Current evidence strongly indicates the former mechanism, since non-hydrolysable ATP analogues are much more effective than ADP and P_i at releasing substrate proteins[16,18]. Further, the addition of ATP to cpn60 leads to the rapid formation of a weak collision complex, followed by a protein structural rearrangement after which ATP is bound tightly[16]. The energy of ATP binding is, therefore, used to drive cpn60 into a state which will release proteins. The system is slowly returned to the tight protein-binding form by hydrolysis of ATP. In keeping with these observations, the association of ATP is seen to cause a co-ordinated reorientation of the subunits in the cpn60 oligomer visible by electron microscope[14].

As expected in an oligomer undergoing a co-ordinated subunit rearrangement, the binding of ATP and its non-hydrolysable analogues is highly co-operative[16]. Such behaviour will co-ordinate the release of the bound protein from all subunits at once to allow folding to progress. The presence of cpn10 in the complex increases the co-operativity still further and some evidence suggests that it is also able to displace protein substrates by direct competition[22]. These factors help explain the requirement for cpn10 in the refolding of very tightly associated proteins.

From the arguments presented thus far, it is clear that the cyclic turnover of ATP switches the chaperonin between weak and tight protein-binding states. We now consider how this cycle is used to improve the efficiency of protein folding. There are two, not necessarily mutually exclusive, models to explain this.

The first is based on the observation that chaperonins suppress the aggregation of protein molecules during folding[12]. The folding of a polypeptide chain is seen to proceed in two stages. In the first few milliseconds the disordered, random coil partially compacts and forms a degree of native secondary structure. This intermediate, or set of intermediates, has been termed the 'molten globule', owing to the core of the molecule being maintained only by weak contacts formed by mobile (or 'liquid'), hydrophobic side-chains. The next step is slower and is described by an interdigitation (or 'solidification') of side-chains and a stabilization of the fully native secondary structure. In oligomeric proteins the folded polypeptide chains must then associate by the

precise pairing of hydrophobic subunit interfaces. In principle, aggregation —
or the disordered and irreversible association of polypeptide chains — could
occur at any stage prior to the formation of the native molecule, but it has been
claimed that chaperonins preferentially bind the molten globule form[13]. This
would increase the yield of native protein by reducing the concentration of
free, partly folded structures which could collide, stick and form irrecoverable
aggregates.

One problem with this view is that binding the intermediate to prevent
aggregation will inevitably reduce the rate of folding because hydrophobic
contacts between the chaperonin surface and the polypeptide must be broken
for folding to proceed. In practice, the chaperonin/ATP-assisted folding reaction usually proceeds at or around the same rate as spontaneous folding, but
with higher yield. It is possible to side-step this criticism by proposing that the
protein is caged within the chaperonin cavity and there is little of the contact
which would otherwise inhibit folding[22]. If this were the case, it is difficult to
see how the substrate protein is attracted into the cavity in the first place and
how the chaperonin can distinguish between native and non-native structures.
In addition, the size of the cavity is too small to encapsulate larger cellular proteins and cryo-electron micrographs of the cpn60–cpn10–ATP complex
revealed the substrate protein and cpn10 are bound on opposite rings of the
cpn60[14].

An alternative mechanism is one in which misfolded structures are unfolded by hsp60 and then released to allow another chance to fold productively[16].
This recycling activity can be likened to taking a solution of unfolded molecules and removing the denaturant. Folding will ensue, and while some
molecules form productive intermediate conformers which are able to fold to
the native structure, other molecules misfold and become kinetically trapped as
unproductive conformers. If the latter class is selectively removed, unfolded
and allowed to fold again, then a further proportion will reach the native structure. Repeated cycles will produce high yields of native protein. Many of the
properties of chaperonins are consistent with this mode of action, which is
summarized in Figure 2. As described earlier, the cycle of ATP binding and
hydrolysis switches the chaperonin between tight and weak protein-binding
states. The interaction energy of the tight-binding form of hsp60 is sufficient
to unfold misfolded structures. On transition to the weak-binding form (upon
ATP binding), the substrate is released to fold again; the chaperonin can,
therefore, act as a recycler of unproductive misfolded structures. In such a
mechanism, misfolded states would be selected over native or well-ordered and
productive intermediates by virtue of a greater exposure of hydrophobic
residues. The fact that such an action can also suppress aggregation of these
poorly folded and 'sticky' forms by holding them on the chaperonin surface
gives the two mechanisms a point of convergence.

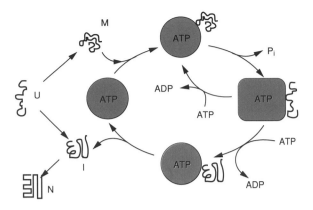

Figure 2. A mechanism for hsp60-assisted protein folding
As in the case of hsp70, ATP stabilizes a weak protein-binding state (round) and hydrolysis returns it to a tight-binding state (square) which can unfold proteins. The reaction steps and the involvement of the co-protein, cpn10, are described in the text. The overall effect is to salvage misfolded structures for recycling and to sequester 'sticky' structures on the chaperonin surface. The fact that many ATP molecules are hydrolysed per protein folded can be accounted for by epicycles in which the bound, unfolded state misfolds and remains on the chaperonin rather than dissociating as a productive intermediate.

TCP1 ring complex: a eukaryotic chaperone in the cytosol

The TCP1 ring complex (TRiC) is constitutively expressed in the cytosol of all mammalian cell types, *Drosophila melanogaster* and yeast[4]. Electron microscopy reveals the structure of the murine and human TCP1 to be a hetero-oligomeric particle of 950 kDa arranged as two rings 12–16 nm in diameter[23]. Each ring is formed by eight subunits and has a large central cavity. Since TRiC has structural similarities with cpn60, it has been suggested that TRiC may act as a molecular chaperone in the cytosol of eukaryotes.

TRiC has been shown to mediate the folding of unfolded β-actin. In the absence of ATP, actin was observed to form a stable complex with the TRiC[24]. The presence of Mg–ATP causes a conformational change in the ring complex which allows the release of the protein substrate and the production of native actin monomers. This kind of mechanism is similar to that observed with the hsp60 chaperonins. However, as yet, no equivalent of the co-chaperonin, cpn10, has been identified for TRiC. It may well be that a co-chaperonin is not needed, since studies comparing the refolding of tubulin by TRiC and the *E. coli* chaperonins demonstrated that TRiC was able to fold tubulin efficiently, while cpn60 was only able to do so in the presence of cpn10. More interestingly, firefly luciferase was unable to be refolded by cpn60, even in the presence of cpn10, whereas purified TRiC was able to fold luciferase efficiently. Electron microscopy revealed that TRiC was associated with hsp70, again raising the possibility that there may be co-operation between the different classes of molecular chaperone. It may well be that TRiC has similar functions to cpn60 in the cytosol of eukaryotes, however, its precise role in protein folding *in vivo* has yet to be discovered.

Hsp90: molecular chaperone or protein regulator?

In eukaryotic cells hsp90s are highly abundant, comprising 1–2% of cytoplasmic protein under normal conditions and rising under conditions of stress[4]. Cytosolic hsp90 and grp94 (ER hsp90) have molecular masses ranging from 82 to 94 kDa in different organisms. Prokaryotes contain a single hsp90 gene, HtpG, encoding a 71.4 kDa protein which has approximately 40% sequence homology with its eukaryotic counterparts.

Walsh and co-workers have purified HtpG and hsp82 (yeast hsp90 homologue) and compared their ATPase activity with that found in purified rat, trypanosomal and human hsp90s[25]. The rates differed greatly, ranging from 150 min^{-1} (trypanosomal) to 0.6 min^{-1} (rat). This ATPase activity could sometimes be stimulated by the binding of substrate proteins. The trypanosomal hsp90 ATPase could be stimulated up to five-fold by the addition of peptides, while the heat-shock factor from HeLa cells enhanced the ATPase of human hsp90 by approximately three-fold. However, the mechanistic role of this stimulation is not yet clear. Hsp90 also undergoes slow autophosphorylation on serine and threonine residues, but phosphorylation has no effect upon the ATPase activity. However, it has been suggested that phosphorylation may regulate the ability of hsp90 to bind its protein targets.

Hsp90s have long been known to associate with a variety of proteins, such as the steroid receptors for oestrogen, progesterone and glucocorticoids, as well as actin and tubulin, casein kinase II, eIF2α kinase and pp60src, the retroviral transforming tyrosine kinase. It has been shown that the inclusion of hsp90 in samples of non-native proteins prevents aggregation and is able to improve the folding efficiency to the native state[26]. In the case of the steroid receptors, hsp90 associates with the protein to hold it in a partially unfolded state unable to bind DNA. On binding the steroid hormone, the receptor dissociates from hsp90 and is then able to bind to specific sites on DNA to regulate transcription. It appears that hsp90 has a multitude of functions conferred by its ability to bind, reversibly, to proteins with exposed hydrophobic surfaces. By switching such proteins on and off, hsp90 may be crucial in the regulation of a wide range of intracellular processes.

Summary

- *Molecular chaperones can be broadly defined as proteins which interact with non-native states of other protein molecules.*
- *This activity is important in the folding of newly synthesized polypeptides and the assembly of multisubunit structures; the maintenance of proteins in unfolded states suitable for translocation across membranes; and the stabilization of inactive forms of proteins which are turned on by cellular signals; and the stabilization of proteins unfolded during cellular stress.*

- *The major chaperone classes are hsp60 (including TCP1), hsp70 and hsp90. All these proteins prevent the aggregation of unfolded proteins and the strength of interaction with their protein substrates is modified by the binding and hydrolysis of ATP.*

- *Hsp70 is a dimeric and ubiquitous protein which binds its substrates in an extended conformation through hydrophobic interactions. It binds to newly synthesized proteins and is required for protein transport. In its ATP-bound state it has a low protein affinity but when the nucleotide is hydrolysed to give the ADP state the affinity is increased. Hsp70 in E. coli (DnaK) is regulated by two co-proteins: DnaJ (of which there are homologues in eukaryotes) stimulates hydrolysis of ATP and GrpE promotes the dissociation of ADP to allow rebinding of ATP. Thus DnaJ promotes the association of substrate proteins and GrpE promotes dissociation.*

- *Hsp60 is a large, tetradecameric protein with a central cavity in which non-native protein structures are proposed to bind. It is essential for the folding of a huge spectrum of unrelated proteins and is present in all biological compartments except the ER. As in hsp70, the binding of ATP stimulates release of the substrate and its hydrolysis restores high binding affinity. It functions in conjunction with a co-protein, cpn10, which enhances its ability to eject proteins during the ATPase cycle. The enhancement of folding yields arises either from the prevention of irreversible aggregation or the ability to unfold misfolded structures and allow further attempts to arrive at the native state.*

- *Proteins of the hsp90 class are found associated with inactive or unstable substrate proteins within the cell, thus preventing their aggregation and/or permitting rapid activation.*

The authors wish to thank the Lister Institute and the Wellcome Trust for supporting their work.

References

1. Anfinsen, C.B. (1973) Principles that govern the folding of protein chains. *Science* **181**, 223–230
2. Creighton, T.E. (ed.) (1992) *Protein Folding*, Freeman, New York
3. Ellis, R.J. & van der Vies, S.M. (1991) Molecular chaperones. *Annu. Rev. Biochem.* **60**, 321–347
4. Hendrick, J.P. & Hartl, F.-U. (1993) Molecular chaperone functions of heat-shock proteins. *Annu. Rev. Biochem.* **62**, 349–384
5. Pelham, H.R.B. (1986) Speculations on the functions of the major heat-shock and glucose-regulated proteins. *Cell* **46**, 959–961
6. Flaherty, K.M., DeLuca-Flaherty, C. & McKay, D.B. (1990) 3-dimensional structure of the ATPase fragment of a 70 K heat shock cognate protein. *Nature* **346**, 623–628
7. Liberek, K., Skowyra, D., Zylicz, M., Johnson, C. & Georgopoulos, C. (1991) The *Escherichia coli* DnaK chaperone, the 70-kDa heat-shock protein eukaryotic equivalent, changes conformation upon ATP hydrolysis, thus triggering its dissociation from a bound target protein. *J. Biol. Chem.* **266**, 14491–14496

8. Palleros, D.R., Reid, K.L., Shi, L., Welch, W.J. & Fink, A.L. (1993) ATP-induced protein hsp70 complex dissociation requires K$^+$ but not ATP hydrolysis. *Nature (London)* **365**, 664–666

9. Flynn, G.C., Rohl, J., Flocco, M.T. & Rothman, J.E. (1991) Peptide-binding specificity of the molecular chaperone BiP. *Nature (London)* **353**, 726–730

10. Langer, T., Lu, C., Echols, H., Flanagan, J., Hayer, M.K. (1992) Successive action of DnaK, DnaJ and GroEL along the pathway of chaperone-mediated protein folding. *Nature (London)* **356**, 683–689

11. Kang, P.J., Ostermann, J., Shilling, J., Neupert, W., Craig, E.A. & Pfanner, N. (1990) Requirement for hsp70 in the mitochondrial matrix for translocation and folding of precursor proteins. *Nature (London)* **348**, 137–143

12. Buchner, J., Schmidt, M., Fuchs, M., Jaenicke, R., Rudolph, R., Schmid, F.X. & Kiefhaber, T. (1991) GroE facilitates refolding of citrate synthase by suppressing aggregation. *Biochemistry* **30**, 1586–1591

13. Martin, J., Langer, T., Boteva, R., Schramel, A., Horwich, A.L. & Hartl, F.-U. (1991) Chaperonin-mediated protein folding at the surface of GroEL through a molten globule-like intermediate. *Nature (London)* **352**, 36–42

14. Saibil, H.R., Zheng, D., Roseman, A.M., Hunter, A.S., Watson, G.M.F., Chen, S., auf der Mauer, A., O'Hara, B.P., Wood, S.P., Mann, N.H., *et al.* (1993) Location of a folding protein and shape changes in GroEL–GroES complexes imaged by cryo-electron microscopy. *Curr. Biol.* **3**, 265–273

15. Schmidt, M., Rutkat, K., Rachel, R., Pfeifer, G., Jaenicke, R., Viitanen, P., Lorimer, G. & Buchner, J. (1994) Symmetric complexes of GroE chaperonins as part of the functional cycle. *Science* **265**, 656–659

16. Jackson, G.S., Staniforth, R.A., Halsall, D.J., Atkinson, T., Holbrook, J.J., Clarke, A.R. & Burston, S.G. (1993) Binding and hydrolysis of nucleotides in the chaperonin catalytic cycle: implications for the mechanism of assisted protein folding. *Biochemistry* **32**, 2554–2563

17. Todd, M.J., Viitanen, P.V. & Lorimer, G.H. (1994) Dynamics of the chaperonin ATPase cycle: implications for facilitated protein folding. *Science* **265**, 659–666

18. Badcoe, I.G., Smith, C.J., Wood, S., Halsall, D.J., Holbrook, J.J., Lund, P. & Clarke, A.R. (1991) Binding of a chaperone to the folding intermediates of lactate dehydrogenase. *Biochemistry* **30**, 9195–9200

19. Gray, T.E. & Fersht, A.R. (1993) Refolding of barnase in the presence of GroE. *J. Mol. Biol.* **232**, 1197–1207

20. Zahn, R., Spitzfaden, C., Ottiger, M., Wüthrich, K. & Plückthun, A. (1994) Destabilization of complete protein secondary structure on binding to the chaperone GroEL. *Nature (London)* **368**, 261–265

21. Okazaki, A., Ikura, T., Kikaido, K. & Kuwajima, K. (1994) The chaperonin GroEL does not recognize apo-α-lactalbumin in the molten globule state. *Nat. Struct. Biol.* **1**, 439–446

22. Martin, J., Mayhew, M., Langer, T. & Hartl, F.-U. (1993) The reaction cycle of GroEL and GroES in chaperonin-assisted protein folding. *Nature (London)* **366**, 228–233

23. Lewis, V.A., Hynes, G.M., Zheng, D., Saibil, H. & Willison, K. (1992) T-complex polypeptide 1 is a subunit of a heteromeric particle in the eukaryotic cytosol. *Nature (London)* **358**, 249–252

24. Gao, Y., Thomas, J.O., Chow, R.L., Lee, G.H. & Cowan, N.J. (1992) A cytoplasmic chaperonin that catalyses β-actin folding. *Cell* **69**, 1043–1050.

25. Nadeau, K., Das, A. & Walsh, C.T. (1993) Hsp90 chaperonins possess ATPase activity and bind heat-shock transcription factors and peptidyl-prolyl isomerases. *J. Biol. Chem.* **268**, 1479–1487

26. Wiech, H., Buchner, J., Zimmerman, R. & Jakob, U. (1992) Hsp90 chaperonins assist protein folding *in vitro*. *Nature (London)* **358**, 169–170

8

Affinity precipitation: a novel approach to protein purification

Jane A. Irwin and Keith F. Tipton

Department of Biochemistry, Trinity College, Dublin 2, Ireland

Introduction

Over the past three decades, many techniques have become available for the purification of proteins. Before the development of suitable media for chromatography, strategies for protein purification included changing the temperature, pH, ionic strength and the precipitation of proteins with the aid of organic solvents, heavy metals or ionic polymers. However, these techniques and ion-exchange chromatographic techniques were not, in general, specific for a particular protein. The technique of affinity chromatography, first developed by Cuatrecasas and co-workers in the late 1960s[1], introduced a purification procedure which could be specific for a given protein. This powerful technique takes advantage of the specific interaction of a biological ligand, such as a substrate, coenzyme, hormone, antibody or nucleic acid, or its synthetic analogue, with its complementary binding site on a protein. This made it possible for the first time to carry out a one-step purification of a protein from a complex mixture.

The development of affinity chromatography gave rise to other, more specialized techniques (for reviews, see Chen[2] and Luong and Nguyen[3]), such as affinity precipitation, affinity partitioning and affinity ultrafiltration. Affinity precipitation, like affinity chromatography, makes use of the specificity of a protein for a ligand. However, instead of binding the ligand to an insoluble support (as in affinity chromatography), followed by elution with a competing ligand or by changing the pH or ionic strength, the protein is precipitated from solution after binding to its ligand. Other contaminants remain in the supernatant and can easily be removed.

There are two main approaches to affinity precipitation. The first of these is the 'bis-ligand' or homobifunctional ligand approach. This involves the synthesis of a bivalent ligand in which two identical ligands are linked via a spacer arm. The bis-ligand can then bind to the ligand-binding sites of two proteins, if the interlinking spacer is long enough. An oligomeric protein can thus bind more than one bis-ligand, and give rise to the formation of a lattice. On reaching a specific size, the cross-linked protein aggregate will precipitate from solution. The protein is then recovered by centrifugation and disaggregation of the complex. Such an approach has the potential advantage of providing a one-step purification that does not involve the use of chromatographic columns or high-speed centrifugation.

In the second approach, the affinity ligand has two functions: to bind the protein and to promote precipitation. Commonly, a ligand specific to the target protein and a precipitation group are attached to a water-soluble polymer. The protein then binds to its ligand and the macroligand can then be precipitated by changing the pH, the temperature or the ionic strength. Alternatively, the ligand which binds the protein can be attached to a carrier molecule which is precipitated by the addition of a third component, such as lectins or metal ions, causing cross-linking and precipitation of the water-soluble polymer. This second technique is also described in the literature as affinity precipitation, but it must be noted that the precipitation is not as a direct consequence of the formation of affinity complexes. As a result, the term 'affinity precipitation' in this context may be something of a misnomer.

Affinity precipitation with bis-ligands

N_2,N_2'-adipodihydrazido-bis-(N^6-carbonylmethyl)-NAD$^+$

Many enzymic reactions involve adenine nucleotide coenzymes and, as a result, techniques of immobilizing these coenzymes, particularly NAD(P)$^+$ and ATP, have been developed for the purposes of affinity chromatography. These immobilization techniques generally involve derivatizing the adenine moiety in such a way that the coenzymic activity is maintained. X-ray crystallographic studies on several enzymes have shown that the exocyclic N^6 position is directed away from the binding site for the adenosine moiety into the medium, so substitution at this position has been a common strategy. As these coenzyme derivatives bind a wide range of dehydrogenases and kinases, and thus are of benefit in purifying numerous enzymes, they have been termed 'general ligands'.

The technique of affinity precipitation was pioneered by Mosbach and colleagues in Sweden in the late 1970s[4]. They synthesized the bifunctional coenzyme derivative N_2,N_2'-adipodihydrazido-bis-(N^6-carbonylmethyl)-NAD$^+$, known as bis-NAD$^+$. It consists of two molecules of N^6-carboxymethyl-

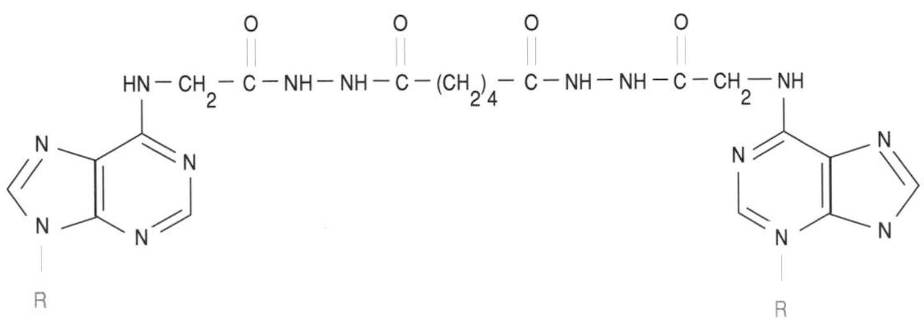

Figure 1. The structure of (bis-NAD⁺), first synthesized by Larsson and Mosbach (1979) as an affinity precipitation ligand for NAD⁺-dependent dehydrogenases
R represents the nicotinamide mononucleotide phosphoribose part of the molecule.

NAD⁺, linked by a spacer of adipic acid dihydrazid (Figure 1). Initial attempts to utilize bis-NAD⁺ for the affinity precipitation of enzymes involved lactate dehydrogenase (LDH). A purified solution of LDH was precipitated — up to 90% yield was obtained — in the presence of pyruvate to promote binding of bis-NAD⁺. Further work in Mosbach's laboratory extended the range of enzymes to which this technique was applicable, to include beef liver glutamate dehydrogenase (GDH) and yeast alcohol dehydrogenase (YADH)[5]. However, the latter enzyme required the presence of salt to enhance precipitation. The recovery of the enzyme from the precipitate was accomplished by the addition of NADH, which displaced the bis-NAD⁺.

Certain criteria have to be fulfilled to allow affinity precipitation to occur. These are as follows: (i) the enzyme has to contain more than one binding site; (ii) the bis-ligand has to have a strong affinity for the enzyme; and (iii) the spacer connecting the two ligands has to be long enough to bridge the distance between two ligand-binding sites on two different molecules.[6]

In addition to point (iii), the ratio of coenzyme derivative to enzyme subunit was also found to be important. If the ratio of bis-coenzyme to enzyme is low, there will be insufficient interlinking to form a lattice. If the ratio of bis-coenzyme to enzyme is too high, each enzyme-binding site may be filled by one end of a bis-coenzyme molecule, but these will not give rise to intermolecular cross-linking. At an optimum ratio of bis-coenzyme to enzyme subunit, cross-linking and subsequent lattice formation occurs. For tetramers, this ratio was found to be approximately unity (1.25 for LDH)[5]. A schematic diagram is shown in Figure 2. The precipitation yield for GDH was found to be high over a wide range of ratios in the presence of glutarate, with up to 70% at an NAD⁺-equivalent/enzyme subunit ratio as low as 0.16 (Figure 3). However, mammalian GDH is a hexamer which exhibits the ability to form aggregates at high concentrations[7] and this may explain this observation.

The ratio close to unity shown in the case of tetramers resembles the behaviour of immunoprecipitation, which displays a similar dependence on the ratio of antibody to antigen for optimum formation of immune complexes, i.e. two antigen molecules per antibody[8]. The similarity between the two

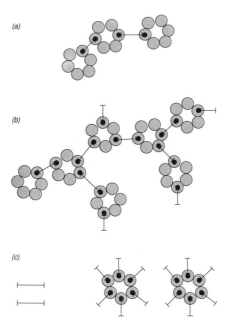

Figure 2. Effect of bis-ligand concentration on affinity precipitation
This is illustrated here with the hexameric enzyme GDH. (*a*) If the ratio of ligand to enzyme sub-unit is low, affinity precipitation will not occur. (*b*) At an optimum ratio (four NAD^+ equivalents per enzyme subunit for GDH, approximately one for YADH and LDH) maximal lattice formation and precipitation will occur. (*c*) When the ratio of NAD^+ equivalents to active sites exceeds the optimum value, a decrease in precipitation occurs because each active site is occupied by a differ-ent bis-NAD^+ molecule. Bis-NAD^+ is represented by a line and glutarate, the substrate analogue for GDH, is shown as a dot. Each bis-NAD^+ molecule comprises two NAD^+ equivalents.

phenomena was exploited by Larsson and Mosbach[5] by carrying out a precipitation–diffusion experiment (Ouchterlony double diffusion) with LDH and bis-NAD^+ in an agarose gel containing pyruvate. They considered that this could have potential clinical applications, e.g. in the detection of abnormal levels of enzymes and other serum proteins, thus avoiding the necessity to raise antibodies to these proteins.

The spacer length of bis-NAD^+ is approximately 1.7 nm, which was found to be long enough to allow easy access simultaneously to the active sites of two dehydrogenase molecules, provided that the sites were not too deeply embed-ded within the enzyme. Other forms of bis-NAD^+ with shorter and longer spacer arms (0.7 nm and 3.2 nm, respectively) have been synthesized[9]. These bis-NAD^+ analogues were used to set up an immobilized two-enzyme system, consisting of LDH and horse liver alcohol dehydrogenase (ADH), which were co-immobilized by glutaraldehyde coupling. One NAD^+ moiety occupied an active site on each enzyme, ensuring that the active sites were positioned against each other, even after removal of the coenzyme analogue. The advan-tage of this was that the diffusion of the product of the first enzyme, i.e. NADH, might be facilitated, since the active site of the other enzyme would, in theory, be nearby.

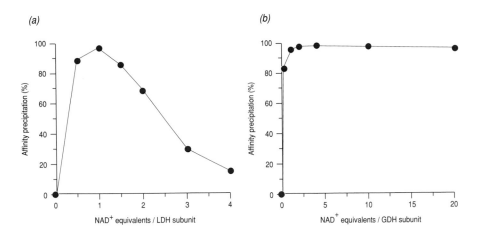

Figure 3. Affinity precipitation of LDH and GDH as a function of bis-NAD$^+$ concentration

(a) Purified bovine muscle LDH (0.5 mg/ml) was affinity precipitated for 16 h at 4 °C in the presence of 28 mM oxalate, 20 mM sodium phosphate buffer, pH 7.4 and varying concentrations of the bis-NAD$^+$ derivative, N,N'-bis(N^6-ethylene-NAD$^+$)pimelamide (J.A. Irwin & K.F. Tipton, unpublished work). One bis-NAD$^+$ molecule comprises two NAD$^+$ equivalents. (b) Purified bovine liver GDH (0.2 mg/ml) was affinity precipitated for 16 h at 4 °C in the presence of 35 mM glutarate, 20 mM sodium phosphate buffer, pH 7.4 and varying concentrations of N,N'-bis[(N^6-ethylene-NAD$^+$)succinamyl] hexamethylene diamine.

These experiments also indirectly demonstrated the feasibility of cross-linking active sites using bifunctional coenzyme derivatives with both relatively short and longer spacer lengths. However, an earlier and analogous study by Green and colleagues on bis-biotinyl diamines and avidin illustrated the importance of spacer lengths in some systems[10]. A series of bis-biotinyl derivatives was synthesized in which the carboxyl groups of the biotin residues were separated by polymethylenediamine spacers. It was found, by electron microscopy, that only one of the two biotin residues could bind to an avidin molecule if the spacer length was less than 1.4 nm. Compounds with longer linkers gave rise to linear polymers of avidin but, when the spacer length exceeded 3.8 nm, the polymers became considerably shorter, suggesting that intramolecular cross-linking was taking place between binding sites on the same avidin molecule.

Another cross-linking study[11], employing the same principle as affinity precipitation with bis-NAD$^+$, was carried out with erythrocytes. These were agglutinated with bifunctional boronic acid derivatives, which interacted with cell-surface glycoproteins.

Bis-NAD$^+$ has been utilized to purify GDH from rat and bovine liver, with a 140-fold and 180-fold purification, respectively[12]. The relative rapidity of purification by this procedure avoided the limited proteolysis which had been observed when using some other procedures. Further work in the same laboratory demonstrated the purification of LDH and the partial purification of isocitrate dehydrogenase from this source[13].

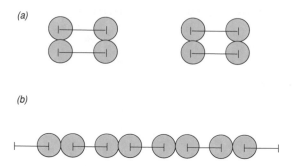

Figure 4. Affinity precipitation of dimeric enzymes
The addition of bis-NAD$^+$ and a substrate analogue to a dimeric dehydrogenase could, in theory, lead to the formation of either dimers of dimers (*a*) or a linear polymer (*b*), which may precipitate out of solution on reaching a sufficiently large size. Bis-NAD$^+$ is represented as a line.

To date this method has not been shown to accomplish the affinity precipitation of dimeric dehydrogenases, such as horse liver ADH. These would be likely to form either dimers of dimers or linear polymers, as shown in Figure 4. Evidence for the formation of the former with horse liver ADH has been found by gel filtration and analytical ultracentrifugation[5], backed up by further studies involving affinity gel filtration[14]. The enzyme was gel-filtered through a Sephadex G-100 column which had been equilibrated with 10 mM pyrazole. The ADH that had been complexed with bis-NAD$^+$ eluted before the uncomplexed enzyme; however, there was no ADH activity in the void volume, which indicated that polymeric aggregates were not present. In such cases, repetitive size-exclusion chromatography or ultrafiltration in the presence and absence of cross-linking can form the basis of a purification procedure.

Affinity precipitation has also been used to look at the possible separation of LDH isoenzymes on the basis of abortive complex formation[14]. The mammalian and avian isoenzymes, found in somatic cells, have been designated H (heart) and M (muscle), as they were originally found to predominate in these organs. The homotetrameric enzymes H$_4$ and M$_4$ have different kinetic, immunochemical, physical and electrophoretic properties, while the intermediate species (H$_3$M, H$_2$M$_2$ and HM$_3$) have intermediate properties depending on their subunit composition. The heart-type isoenzyme (H$_4$) forms an abortive complex with NAD$^+$ and pyruvate, whereas the M$_4$ isoenzyme does not. Bis-NAD$^+$ gave rise to affinity precipitation of the H$_4$ isoenzyme in the presence of oxamate, a competitive inhibitor with respect to pyruvate, but the M$_4$ isoenzyme did not precipitate under these conditions.

A more recent study (J.A. Irwin and K.F. Tipton, unpublished work) has shown that intermediate isoenzymes of LDH can also be affinity precipitated in the presence of bis-NAD$^+$ and oxamate. Several different forms of bis-NAD$^+$ were synthesized by condensing N^6-(2-aminoethyl)NAD$^+$ and four different carboxylic acids in a carbodiimide-mediated reaction (Figure 5). The H_4, H_3M and H_2M_2 isoenzymes were affinity precipitated, both from a mixture of all the enzymes and in the purified forms. However, small amounts of the HM_3 and M_4 isoenzymes were also shown to affinity precipitate (as shown by starch gel electrophoresis) possibly due to entrapment in the lattices of cross-linked enzyme as they formed. These bis-NAD$^+$ derivatives also precipitated GDH and YADH, which was purified to homogeneity from a yeast lysate, as shown by PAGE. A bacterial alanine dehydrogenase and glyceraldehyde-3-phosphate dehydrogenase from yeast and rabbit muscle were found to be refractory to precipitation. All forms of bis-NAD$^+$ synthesized were reduced by YADH and ethanol, and these bis-NADH derivatives affinity precipitated purified solutions of YADH, GDH and LDH in the absence of a substrate analogue.

The 'locking-on' effect

Since NAD$^+$, NADP$^+$ and ATP are 'general ligands', affinity chromatography using these ligands has the drawback that elution with a general competitive counter-ligand, such as a salt, or the coenzyme itself will lead to the elution of more than one enzyme. Similarly, bis-NAD$^+$ can be described as a general ligand, since it also binds to more than one enzyme.

Figure 5. Synthesis and structure of bis-NAD$^+$ derivatives

Two molecules of N^6-(2-aminoethyl)NAD$^+$ are condensed with one molecule of a dicarboxylic acid ($n = 3$, 4 or 5 for glutarate, adipate and pimelate, respectively) in the presence of 1-ethyl-3,3-dimethyl aminopropyl carbodiimide (EDC) at pH 5.5–7.0. R represents the nicotinamide mononucleotide phosphoribose portion of the molecule.

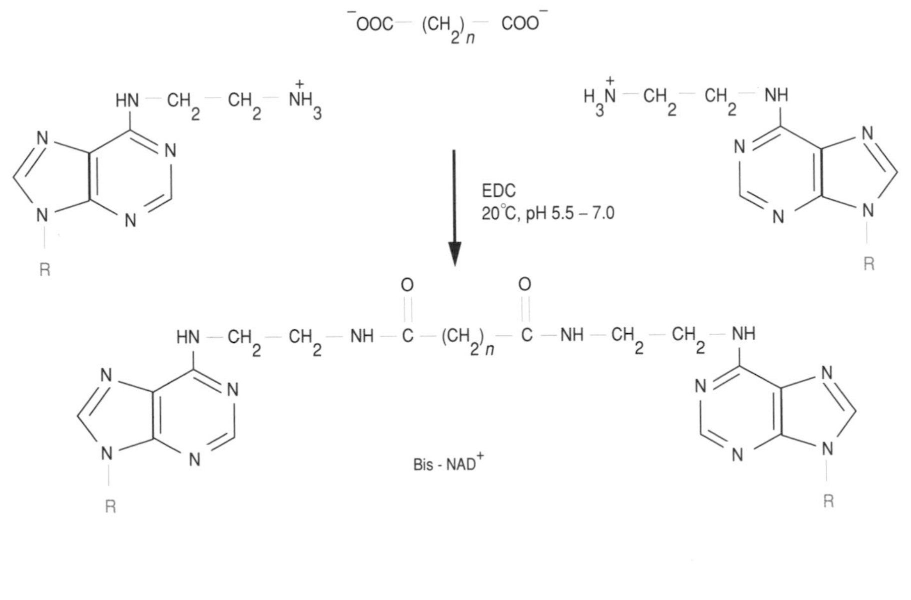

This problem was circumvented by O'Carra and colleagues[16] by exploiting what they described as the 'locking-on' effect. This involved increasing the strength of enzyme binding to an immobilized ligand by adding analogues of substrates specific for the enzyme of interest. The strength of adsorption of LDH, for example, on to an immobilized NAD^+ derivative was increased by adding oxalate, a structural analogue of lactate, to the irrigating buffer. The omission of oxalate from the eluting buffer weakened the binding, allowing the LDH to be readily eluted from the column.

The locking-on mechanism does not occur with all enzymes. Enzyme reaction mechanisms can, in general, be divided into two main categories: sequential and non-sequential (Ping-Pong). In the former category, all substrates must bind to the enzyme before products are released. Furthermore, sequential pathways can be either ordered sequential (where the order or substrate addition is fixed) or random sequential, where the substrates can add on in random order. Most coenzyme-dependent enzymes, including most dehydrogenases, exhibit ordered sequential mechanisms, with binding of the co-enzyme preceding the binding of the second substrate.

This affects the locking-on effect inasmuch as it will only apply if the enzyme has an ordered mechanism, in which the 'leading' ligand (in this case the coenzyme) binds to the enzyme before the secondary ligand or substrate. In such cases an unreactive, competitive analogue of the second substrate will displace the coenzyme-binding equilibrium by ternary-complex formation, thus effectively increasing the strength of binding. However, a form of the locking-on effect can also occur in the case of a random sequential mechanism, if the equilibrium of the reaction under the conditions used is such that the binding of the second substrate favours the binding of the coenzyme.

These mechanisms are illustrated below. In the case of an ordered sequential mechanism, the coenzyme (A) binds to the enzyme before the analogue of the second substrate (B).

$$E \; \xrightleftharpoons{\;K_A\;} \; EA \; \xrightleftharpoons{\;K_B\;} \; EAB$$

If the second ligand (B) is at a saturating concentration, the enzyme will be locked into the ternary complex. For a random sequential mechanism, either A or B may bind first.

$$
\begin{array}{ccc}
 & EA & \\
{}^{K_A}\nearrow & & \searrow {}^{K'_A} \\
E & & EAB \\
{}_{K_B}\searrow & & \nearrow {}_{K'_B} \\
 & EB &
\end{array}
$$

For the random order mechanism,

$$EAB = \frac{[E]_{total}}{1 + \dfrac{K_A K'_B}{[A][B]} + \dfrac{K'_B}{[B]} + \dfrac{K'_A}{[A]}}$$

If $[B] \to \infty$, i.e. B is saturating, this reduces to

$$EAB = \frac{[E]_{total}}{1 + \dfrac{K'_A}{[A]}}$$

In an ordered sequential mechanism, the equation becomes

$$EAB = \frac{[E]_{total}}{1 + \dfrac{K_A K'_B}{[A][B]} + \dfrac{K'_B}{[B]}}$$

and thus, if $[B] \to \infty$, this reduces to $EAB = [E]_{total}$ and the enzyme is said to be 'locked' into the ternary complex.

It can be seen from the equation that the situation depends on the value of K'_A. The lower the value of K'_A, the more enzyme is present as the ternary complex, provided that B is saturating. In essence, locking-on only occurs if $K'_A < K_A$. As a result, the substrate concentrations and the values of the equilibrium constants will determine whether or not an enzyme displaying this reaction mechanism will form a ternary complex in the appropriate order to allow locking-on to occur.

In the context of affinity precipitation with bis-NAD$^+$, this concept has a two-fold relevance. First, it can be exploited to give rise to the specific precipitation of a particular enzyme, by a competitive analogue of the oxidizable substrate or, perhaps, the reducible substrate, to form an abortive complex. Addition of the oxidizable substrate itself would give rise to catalytic conversion, an unwanted effect as the bis-NADH thus formed might then dissociate from the active site and no cross-linking would occur.

The system allows the removal of a particular dehydrogenase from a crude extract; for example, the addition of bis-NAD$^+$ and glutarate gives rise to the selective precipitation of GDH from a bovine liver extract[12], even in the presence of other NAD$^+$-dependent dehydrogenases. The formation of a ternary complex between the enzyme, the NAD$^+$ moiety and the substrate or its analogue is the key to conferring specificity on bis-NAD$^+$, enabling it to 'pull' one dehydrogenase out of a mixture containing several.

A second point to note is that the addition of low concentrations of bis-NAD^+ alone to a crude extract will not give rise to affinity precipitation. For example, low concentrations of bis-NAD^+ will not precipitate LDH in the absence of pyruvate or oxalate; this is not surprising, since the binary complex between bis-NAD^+ and LDH is weak and has a dissociation constant (K_d) for NAD^+ of 3×10^{-4} M (see reference 17). Adding a substrate analogue strengthens this binding; for example, Theorell and Yonetani found that NAD^+ bind 1600 times more tightly to horse liver ADH in the presence of pyrazole than in its absence[18]. The addition of NADH to a cross-linked aggregate leads to its disaggregation, because NADH binds, on average, one order of magnitude more tightly to the active sites of many dehydrogenases than does NAD^+ and so displaces the bound NAD^+ if the second substrate or its analogue is not saturating[19].

Bis-ATP

A bis-ATP derivative, N_2,N_2'-(adipodihydrazido)-bis-(N^6-carbonylmethyl)-ATP, has also been synthesized, using N^6-carboxymethyl-ATP as starting material. It has been used to purify bovine heart phosphofructokinase, which precipitates both in purified form and from a crude tissue extract[20]. Precipitation in this case does not involve the simple locking-on effect, discussed above, but instead utilizes citrate to assist in precipitation. ATP acts as both a substrate and an allosteric inhibitor of this enzyme, and the allosteric inhibition is potentiated by citrate, which is also an allosteric inhibitor of phosphofructokinase. However, to date there have been no reports of the precipitation of any other enzymes by bis-ATP. The usefulness of bis-ATP has been limited by its instability, since it does not remain stable for longer than a week, possibly due to phosphate-catalysed cleavage of the dihydrazido-groups in the spacer arm. Clearly, the development of other, more stable, bis-ATP derivatives would be helpful in extending this method to the purification of kinases.

Triazine dyes

One common method now employed for the large-scale purification of proteins is dye–ligand chromatography. Reactive dyes interact with many proteins, and they are easily bound to absorbent matrices, which makes them useful as pseudo-affinity ligands. Not only this, but they are available in bulk at relatively low cost and are chemically stable. For these reasons, they have been investigated, albeit not widely, as affinity-precipitation ligands[21–26].

One of the first affinity-precipitation studies was performed using the bifunctional triazine dye Procion Red HE-3B (Figure 6). This was found to precipitate plasminogen from plasma, and this was incorporated in a purification protocol for plasminogen[25]. Maximal precipitation was obtained above a dye:plasminogen molar ratio of 6 and was pH dependent, with optimum pH at or below pH 5.0. However, Pearson observed a similar effect when simpler

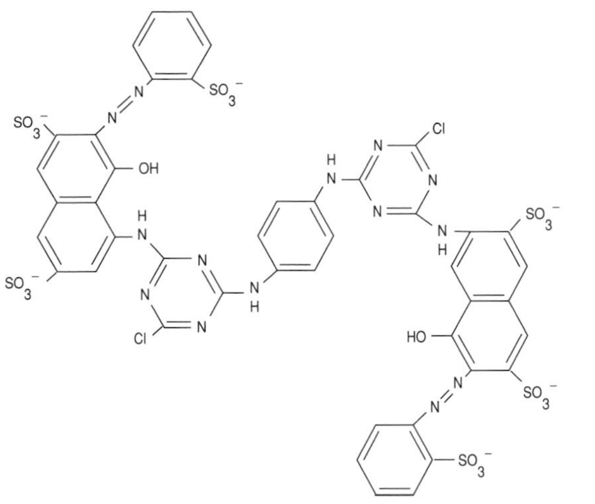

Figure 6. Structure of the triazine dye Procion Red HE-3B (also available under the name C.I. Reactive Red 120)
This bifunctional dye was found to precipitate plasminogen from plasma[25].

triazine dyes were added to solutions of rabbit muscle LDH and bovine liver GDH below pH 6.0[21].

However, these processes were not attributed to selective interactions with these proteins, but rather to the interaction between negatively charged dyes with the positively charged surfaces of proteins, forming insoluble aggregates[25]. The resulting precipitation could not be described as affinity precipitation as such, and attempts were made to devise suitable affinity ligands based on Cibacron Blue F3G-A, which has known affinity for a number of NAD$^+$-dependent dehydrogenases (see Figure 7 for its structure). A bis-derivative of Cibacron Blue F3G-A was synthesized[22], by coupling two dye molecules through sulphonate groups with carbodiimide to form the sulphonamide derivative. A 90% precipitation yield was obtained for LDH, whereas BSA and chymosan gave precipitation yields of only 50% and 20% respectively. Another bis-derivative of this dye was synthesized by reacting it with 6-aminohexyl Cibacron Blue F3G-A, to give a bifunctional dye derivative. It specifically precipitated rabbit muscle LDH in the presence of transition metal ions, such as Co^{2+} and Ni^{2+}, which enhanced the affinity of the dye for the enzyme. Precipitation was inhibited by the addition of NADH or NAD$^+$, and the precipitated enzyme could be solubilized by adding NADH, NAD$^+$-pyruvate or EDTA[23].

Pearson[21] drew up a list of criteria to characterize true affinity precipitation by such compounds, as distinct from precipitation due to non-specific electrostatic interactions. These were as follows.

- Precipitation should depend upon the dye/enzyme ratio, with maximum precipitation at a ratio of one dye-binding equivalent per enzyme-binding site.

- Precipitation should be reversed upon the addition of a compound which competes with the bifunctional ligand for the dye-binding site.

- The bifunctional dye derivative, but not closely related monofunctional dyes, should promote precipitation.

- The alteration of factors, such as pH, temperature or the ionic strength of the medium, should only affect precipitation inasmuch as they affect the interaction between the dye-ligand and the active site.

- Complete recovery of enzyme activity upon resolubilization should be possible, to show that precipitation has not been induced due to gross structural changes in the protein structure.

Two of the examples cited above[22,25] showed little evidence of conforming to these criteria (although the example in reference 23 met some of them). However, affinity precipitation with bis-NAD⁺ does conform if dye is replaced in this context with bis-NAD⁺.

A later study[24] showed that a monofunctional synthetic analogue of the anthraquinone dye, Procion Blue H-B, gave rise to the affinity precipitation of LDH in a selective fashion and in accordance with the criteria laid down above. This derivative was a *p*-methoxylated *o*-sulphonate isomer of Procion Blue H-B (see Figure 7 for structure). Precipitation yields of up to 95% were obtained for solutions of the purified enzyme, in buffers of low ionic strength, and the precipitation displayed little pH dependence over the pH range 6.5–9.0. The complex was resolubilized by the addition of 95 µM NADH. The

Figure 7. Triazine dye structures
(*a*) Cibacron Blue 3G-A (also commercially available under the names Procion Blue H-B or C.I. Reactive Blue 2). (*b*) The methoxylated *p*-sulphonated isomer of Procion Blue H-B, used by Pearson and co-workers to affinity precipitate rabbit muscle LDH from a crude supernatant[26].

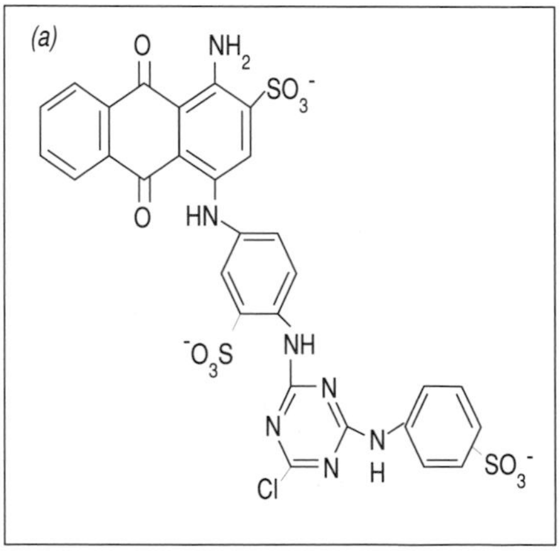

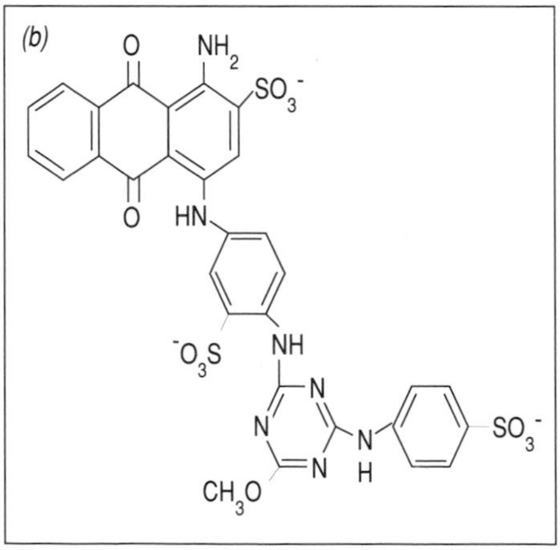

enzyme was also precipitated from a crude rabbit muscle extract, with 97% recovery and a six-fold purification, as judged by the increase in specific activity. Further work[26] indicated that this technique was applicable on a large scale, with a 21-fold purification and a 60% overall recovery of this enzyme from the same source. However, the technique was only applicable to this enzyme, and no precipitation was obtained when the dye was added to a solution of the porcine heart enzyme. It was suggested that the anthraquinone ring occupied one enzyme-binding site in the complex, while the terminal methoxylated triazine and p-aminobenzene sulphonate rings formed a cross-link to a similar binding site on a subunit of a nearby enzyme molecule[24]. The 2:1 molar ratio of enzyme subunit to dye is consistent with such a mechanism. In none of these examples, however, was the locking-on effect exploited.

Thus this technique appears to be specific for the LDH from rabbit muscle and, so far, no evidence has been presented for true affinity precipitation of other enzymes using this mode of affinity precipitation. Notwithstanding its advantages in terms of simplicity, specificity and cost-effectiveness, it appears to be singularly lacking in versatility. In addition to this, care must also be taken to ensure that the free dye does not inactivate the protein of interest, as a number of reactive dichlorotriazine dyes have been found to specifically and irreversibly inactivate yeast hexokinase, yeast glucose-6-phosphate dehydrogenase and pig heart LDH[27].

Metal ion affinity precipitation

Immobilized ion-affinity chromatography has been a recognized technique in protein purification for some time. It exploits the affinity between metal ions and metal-coordinating residues on protein surfaces, such as histidine. However, its application to affinity precipitation, at least using the bis-ligand approach, has been limited to only a few studies. An early study[28] was carried out with bis-copper chelates to cross-link proteins containing multiple surface-accessible histidine residues. Two bis-chelates were synthesized: a short-chain chelate, consisting of two Cu^{2+} ions chelated by a molecule of EGTA and a long-chain chelate, PEG–Cu(II)$_2$, which comprised Cu^{2+} ions chelated by molecules of iminodiacetic acid, immobilized on each end of a molecule of polyethylene glycol (PEG 20000). The postulated structure of the bis-chelate is given in Figure 8. Both bis-chelates were effective in precipitating human haemoglobin and sperm-whale myoglobin, which have 26 and six multiple surface-active histidine residues, respectively. Human haemoglobin was found to give a 100% precipitation yield at a Cu^{2+}/surface-accessible histidine ratio of unity at pH 8.0 when either of the bis-chelates was used; whereas the myoglobin, with fewer accessible histidine residues, gave only a 10% precipitation yield under these conditions. The higher molecular mass bis-chelate was found to be a more effective precipitant of haemoglobin than the Cu(II)$_2$–EGTA complex, inasmuch as it exhibited higher precipitation at lower ratios of Cu^{2+} to available histidine.

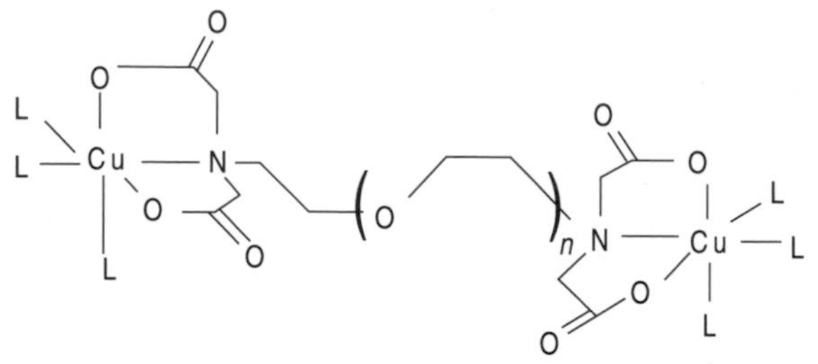

Figure 8. Postulated structure of Cu(II)$_2$EGTA and Cu(II)$_2$–polyethylene glycol–(IDA)$_2$ complexes
IDA is iminodiacetic acid[28].

The process was affected by pH; decreasing the pH from 8.0 to 5.5 gave rise to a decrease in the amount of protein which could be precipitated at a given concentration of Cu^{2+}. When an excess of bis-chelate was used, a 1:1 ratio of Cu^{2+} to surface-accessible histidines was found in the precipitate. Free $CuSO_4$ also caused precipitation, similar to that given by Cu(II)$_2$–EGTA, as a single Cu^{2+} ion cross-linked the proteins in a similar fashion to the Cu(II)$_2$–EGTA chelate. Horse heart cytochrome c, which has only one surface-accessible histidine, was not precipitated by any of the bis-chelates used.

The studies described above were only carried out with pure protein and no attempts were made to dissociate the protein from the complex. However, the development of recombinant DNA technology has also provided a use for metal ion affinity precipitation. Metal ion affinity chromatography had been exploited to purify recombinant proteins and, following from this, Lilius and co-workers[29] chose galactose dehydrogenase as a prototype target protein to investigate this form of affinity precipitation. A DNA fragment encoding five histidine residues was fused to the 3′ terminal end of the galactose dehydrogenase gene from *Pseudomonas fluorescens*, which was then expressed in *Escherichia coli*. This functioned as an affinity tail and, since the enzyme is a homodimer, each enzyme molecule carried two pentahistidine affinity tails. A bis-metal chelate complex, EGTA(Zn)$_2$, bound two galactose dehydrogenase subunits, giving rise to the formation of long chains of linear polymers (see Figure 9). Precipitation yields of up to 90% were obtained at a concentration of 10 mM metal chelate complex. A reference system containing native galactose dehydrogenase gave rise to no precipitation, at the same concentration of the bis-chelate. The complex was dissolved by the addition of the chelating agent EDTA to recover the zinc, and the native enzyme was then formed by digestion of the pentahistidine affinity tail with carboxypeptidase A.

The advantage of this approach to affinity precipitation is its versatility. By fusing an affinity tail to a protein of interest, this method of affinity precipitation can be extended to other recombinant proteins. Furthermore, bis-metal chelates, such as EGTA(Zn)$_2$, are chemically stable and relatively inexpensive.

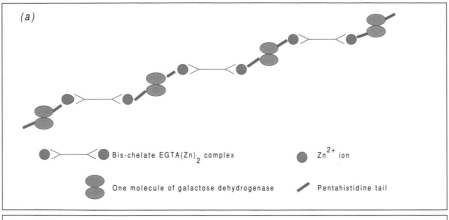

Figure 9. Schematic diagram of affinity precipitation of recombinant galactose dehydrogenase
(a) Possible linear polymeric structures formed between dimeric recombinant GDH molecules, each of which carries one pentahistidine affinity tail per subunit, and the bis-chelate EGTA(Zn^{2+}) affinity precipitating agent. (b) Structure of the bis-chelate EGTA(Zn)$_2$ affinity precipitation agent. Redrawn from reference 29.

However, it is unlikely that this method could be used alone as a one-step purification of a non-recombinant protein, as it would be too non-specific.

Affinity precipitation with heterobifunctional ligands

A second type of affinity precipitation can be described as the heterobifunctional approach. This technique is performed using a macromolecule with two separate functions. One function is to bind to the desired protein, while the second gives rise to the precipitation of the complex. The precipitation of the macroligand is not due to the formation of affinity complexes and the precipitation does not depend on the concentration of the protein of interest. The yield of the protein obtained increases in proportion to the concentration of the macroligand added, in contrast to the situation with bis-ligands, in which the yield usually decreases once the optimum concentration has been exceeded. Furthermore, such systems can be applied not only to the purification of multimeric proteins, but to that of monomers with equal facility. Precipitation of the entire complex was generally accomplished by changing the pH, temperature or ionic strength of the system.

The first example of this approach was demonstrated by Schneider and co-workers[30]. They synthesized a water-soluble polyacrylamide, bearing *N*-acryloyl-*p*-aminobenzamidine and *N*-acryloyl-*p*-aminobenzoic acid moieties.

This was used to purify trypsin from beef pancreas. The *p*-aminobenzamidine moiety bound the trypsin, whereas the *p*-aminobenzoic acid gave rise to precipitation at low pH values, at which its carboxyl group was protonated. The affinity polymer was added directly to a crude extract under conditions favouring enzyme binding, and the pH was then reduced to 3. On removal of the supernatant, the enzyme was eluted from the polymer giving a 90% yield and the polymer was then recycled. The amount of polymer added was low (0.1–0.5%) and the average polymer loss per cycle was only about 1%.

Since then, a range of polymers have been used to promote precipitation. One of these is chitosan, which is composed of partly deacylated chitin. It contains many residues of *N*-acetyl-D-glucosamine and precipitates from solution at high pH values when its amino groups are uncharged. This polymer was used in two studies to purify trypsin and wheat-germ agglutinin, respectively[31,32]. Soybean trypsin inhibitor was coupled to chitin, to serve as a ligand for trypsin. The complex was added to solutions containing trypsin at pH 5.5 and the pH was increased to 8–9 to effect precipitation of the trypsin–soybean trypsin inhibitor–chitosan complex. After pH adjustment to 2.5 to dissociate the trypsin, the derivatized chitosan and trypsin were separated by gel filtration. The enzyme was purified, giving a 93% yield[31]. However, the inclusion of a gel filtration step does lengthen the procedure. Wheat-germ agglutinin, a lectin, was bound directly to the *N*-acetyl-D-glucosamine residues of chitosan. The complex was precipitated at alkaline pH and the precipitate was obtained by dispersed gas flotation[32].

Another polymer with a different pH-solubility profile, hydroxymethylcellulose acetate succinate, has been used as a ligand carrier. It is soluble above pH 5 and insoluble below pH 4.5, and it has been coupled to human immunoglobulin G (IgG) as an affinity ligand to purify staphylococcal protein A. Binding of protein was carried out at pH 7.2, precipitation at pH 4.2, and elution of protein A at pH 2.5[33]. This approach could probably be modified for the purification of other proteins, although its usefulness would be limited by the pH tolerance of the protein of interest.

Precipitation by varying the temperature and ionic strength has been developed using water-soluble synthetic co-polymers of *N*-isopropyl acrylamide (NIPAM). Nguyen and Luong[34] co-polymerized this material with *N*-acryloxysuccinimide or glycidyl methacrylate. This gave rise to the production of ligands with reactive succinimide or epoxy groups, which were reacted with IgG and *p*-aminobenzamidine, thereby synthesizing affinity macroligands for the binding of protein A and trypsin, respectively. The polymer complexes were precipitated by increasing the temperature to 42 °C in the case of the *N*-acryloxysuccinimide derivative and to 34 °C in the case of the other derivative, or by the addition of ammonium sulphate (38 g/l and 27 g/l, respectively). Increasing the temperature above 34 °C gave rise to the recovery of trypsin, an 82% yield was obtained from mixtures of trypsin and chymotrypsin, with 95% recovery of the macroligand.

Protein A was also conjugated to NIPAM, to form a conjugate which binds human IgG. The complex was dissociated by heating it above the lower critical solution temperature of the complex[35]. Conjugates of polymerized NIPAM and a monoclonal antibody have also been synthesized. This system was not utilized to carry out affinity precipitation, but rather to develop a novel immunoassay, as the conjugate was precipitated upon antigen binding[36].

Other polysaccharides besides chitosan have been employed as polymers in affinity precipitation systems. These include dextran[37], galactomannan[38] and alginate[39,40]. LDH from pig heart was bound to Blue Dextran via its attached Cibacron Blue residues. The complex was precipitated by the addition of concanavalin A, a multivalent lectin which cross-linked the dextran and led to the precipitation of the entire complex. The ratio of concanavalin A to Blue Dextran was found to be critical; efficient precipitation occurred only when the ratio of concanavalin A to Blue Dextran was 4.5 (w/w). The percentage LDH obtained was identical to that of Blue Dextran removed and the enzyme was desorbed by increasing the ionic strength, with complete dissociation occurring at 1.5 M KCl. Concanavalin A and Blue Dextran were dissociated by adding glucose, which acted as a competitive ligand with respect to concanavalin A[37].

IgG was recovered from serum using protein A covalently bound to galactomannan, followed by non-covalent cross-linking of the polymer with potassium borate to induce precipitation. IgG was dissociated from the complex by the addition of potassium thiocyanate and the galactomannan–protein A complex was disaggregated by decreasing the pH of the solution to 4. The IgG obtained was found to be 90% pure by PAGE[38]. Similarly, Linné and co-workers affinity precipitated trypsin[38] by means of soybean trypsin inhibitor bound to alginate, a naturally occurring co-polymer of mannuronic and guluronic acid units, isolated from marine algae. Calcium ions were used to precipitate the complex, since divalent or trivalent ions induce alginate precipitation, primarily by interacting with the guluronic acid residues. The precipitate was dissolved by adding EDTA and dissociation of the trypsin was carried out either by decreasing the pH, through addition of HCl, or by displacement with arginine. At least 70% of the trypsin was found to precipitate when both crude and pure trypsin solutions were used. However, the ligand carrier was highly charged, which also induced co-precipitation by the non-specific adsorption of other proteins. Alginate has also been used recently in the affinity precipitation of endo-polygalacturonase[40].

These systems[37,39] were slightly more complex than the systems, such as those using polymerized NIPAM, in which the secondary function of the ligand is used to promote precipitation, e.g. by pH or temperature changes. The precipitation of the target molecule was an indirect process, and depended on the ratio of the macroligand to the cross-linking agent. In this respect, they bear more resemblance to the systems in which homobifunctional ligands were employed.

A more recent study[41] combined the processes of aqueous two-phase extraction and affinity precipitation. Aqueous two-phase separation systems are generally used for the partitioning of enzymes during protein purification and separating proteins from cellular debris. Such systems are generally created by mixing solutions of PEG and dextran or PEG and salts, e.g. potassium phosphate or ammonium sulphate. Human IgG was coupled to Eudragit S100, an enteric coating polymer which is insoluble at low pH but soluble at neutral to weakly alkaline pH. This two-phase extraction process does not always separate a protein specifically from a crude mixture. To enhance the specificity of the system, IgG was coupled to the polymer, to bind recombinant protein A. This remained in the top phase. The top phase was removed, its pH was adjusted to pH 4.5, the precipitate was collected by centrifugation and the pellet was resuspended at pH 2.5 to elute protein A. Recovery of protein A varied from 55–80%, with a 26-fold purification, the lower value being obtained when the system was overloaded.

Conclusion

Affinity precipitation is a quite simple technique which has been shown to be applicable to the purification of a variety of proteins. True affinity precipitation, using bis-ligands such as bis-NAD$^+$, gives rise to precipitation as a direct consequence of the formation of affinity complexes. Specificity is conferred by the addition of substrate analogues to form ternary complexes with the protein and ligand, thus strengthening ligand binding. The advantage of this approach is that it can be used in crude solutions on a large scale to give rise to a one-step purification; however, its widespread use has been limited by the number of proteins for which the method works, the high cost of the ligands and their susceptibility to degradation in crude extracts. To predict whether the approach would work for a given protein, it would be necessary to know its kinetic mechanism, oligomeric structure and the accessibility of its binding sites to the bifunctional ligands to be used.

Triazine dyes and metal ion bis-ligands are relatively inexpensive, widely available, and, in the case of metal ion bis-ligands, easy to synthesize, but they are less specific than bis-coenzymes. However, it is likely that in the future metal ion bis-ligands will be more widely used in the purification of recombinant proteins, as the technique might, in theory, be adapted to a wide range of recombinant proteins by the inclusion of affinity tails.

The development of the heterobifunctional approach has been limited by the availability of affinity polymers which can be affinity precipitated, particularly under conditions which are not injurious to the protein of interest, as many proteins are denatured by extremes of pH or temperature.

The requirements for simple procedures for purifying native and recombinant proteins do not diminish in importance. Despite the drawbacks discussed above, affinity precipitation is attractive as a technique for purifying proteins on a large scale; it can be carried out relatively quickly as a batch step and can often be performed in the absence of any chromatographic steps, whereas a conventional purification might require several chromatographic steps. It also gives rise to high yields of the protein of interest (sometimes over 90%) and, for these reasons, purification by affinity precipitation will probably become more widely used in the future.

References

1. Cuatrecasas, P., Wilchek, M. & Anfinsen, C.B. (1968) Selective enzyme purification by affinity chromatography. *Proc. Natl. Acad. Sci. U.S.A.* **61**, 636–643

2. Chen, J.-P. (1990) Novel affinity based processes for protein purification. *J. Ferment. Bioeng.* **70**, 199–209

3. Luong, J.H.T. & Nguyen, A.-L. (1992) Novel separations based on affinity interactions. *Adv. Biochem. Eng. Biotechnol.* **47**, 138–158

4. Larsson, P.-O. & Mosbach, K. (1979) Affinity precipitation of enzymes. *FEBS Lett.* **98**, 333–338

5. Flygare, S., Griffin, T., Larsson, P.-O. & Mosbach, K. (1983) Affinity precipitation of dehydrogenases. *Anal. Biochem.* **133**, 409–416

6. Larsson, P.-O. & Mosbach, K. (1981) Novel affinity techniques. *Biochem. Soc. Trans.* **9**, 285–287

7. Smith, E.L., Austen, B.M., Blumenthal, K.M. & Nyc, J.F. (1975) Glutamate dehydrogenases, in *The Enzymes*, vol. 11 (Boyer, P.D., ed.), pp. 294–368, Academic Press, New York

8. Feinstein, A. & Rowe, A.J. (1965) Molecular mechanism of formation of an antigen–antibody complex. *Nature (London)* **205**, 147–149

9. Månsson, M.-O., Siegbahn, N. & Mosbach, K. (1983) Site-to-site directed immobilization of enzymes with bis-NAD analogues. *Proc. Natl. Acad. Sci. U.S.A.* **80**, 1487–1491

10. Green, N.M., Konieczny, L., Toms, E.J. & Valentine, R.C. (1971) The use of bifunctional biotinyl compounds to determine the arrangement of subunits in avidin. *Biochem. J.* **125**, 781–791

11. Burnett, T.J., Peebles, H.C. & Hageman, J.H. (1980) Synthesis of a fluorescent boronic acid which reversibly binds to cell walls and a diboronic acid which agglutinates erythrocytes. *Biochem. Biophys. Res. Commun.* **96**, 157–162

12. Graham. L.D., Griffin, T.O., Beatty, R.E., McCarthy, A.D. & Tipton, K.F. (1985) Purification of liver glutamate dehydrogenase by affinity precipitation and studies on its denaturation. *Biochim. Biophys. Acta* **828**, 266–269

13. Beattie, R.E., Graham, L.D., Griffin, T.O. & Tipton, K.F. (1985) Purification of NAD^+-dependent dehydrogenases by affinity precipitation with adipo-N_2,N_2'-dihydrazido-bis-(N^6-carboxymethyl-NAD^+) (bis-NAD^+) *Biochem. Soc. Trans.* **12**, 433

14. Buchanan, M., O'Dea, C.D., Griffin, T.O. & Tipton, K.F. (1989) Reversible cross-linking of alcohol and lactate dehydrogenases with the bifunctional reagent N_2,N_2'-adipodihydrazido-bis-(N^6-carboxymethyl-NAD^+). *Biochem. Soc. Trans.* **17**, 422

15. Reference deleted

16. O'Carra, P. (1978) Theory and practice of affinity chromatography, in *Chromatography of Synthetic and Biological Polymers Vol. 21*, (Epton, R., ed.), pp. 131–158, Ellis Horwood, London

17. Larsson, P.-O., Flygare, S. & Mosbach, K. (1984) Affinity precipitation of dehydrogenases. *Methods Enzymol.* **104**, 364–369

18. Theorell, H. & Yonetani, T. (1963) Liver alcohol dehydrogenase–DPN–pyrazole complex: a model of a ternary intermediate in the enzyme reaction. *Biochem. Z.* **338**, 537–553

19. Dalziel, K. (1975) Kinetics and mechanism of nicotinamide-nucleotide-linked dehydrogenases, in *The Enzymes*, vol. 11 (Boyer, P.D., ed.), pp. 1–61, Academic Press, New York

20. Beattie, R.E., Buchanan, M. & Tipton, K.F. (1987) The synthesis of N_2,N_2'-adipodihydrazido-bis-(N^6-carboxymethyl-ATP) and its use in the purification of phosphofructokinase. *Biochem. Soc. Trans.* **15**, 1043–1044

21. Pearson, J.C. (1987) Fractional protein precipitation using triazine dyes, in *Reactive Dyes In Protein and Enzyme Technology* (Clonis, Y.D., Atkinson, A., Burton, C.J. & Lowe, C.R., eds.), pp. 187–191, Macmillan Press, New York

22. Hayet, M. & Vijayalakshmi, M.A. (1986) Affinity precipitation of proteins using bis-dyes. *J. Chromatogr.* **376**, 157–161

23. Lowe, C.R. & Pearson, J.C. (1983) Bio-mimetic dyes, in *Affinity Chromatography and Biological Recognition*, (Chaiken, I.M., Wilchek, M. & Parikh, I., eds.), pp. 421–432, Academic Press, London

24. Pearson, J.C., Burton, S.J. & Lowe, C.R. (1986) Affinity precipitation of lactate dehydrogenase with a triazine dye derivative: selective precipitation of rabbit muscle lactate dehydrogenase with a Procion Blue H-B analog. *Anal. Biochem.* **158**, 382–389

25. Bertrand, O., Cochet., S., Kroviarski, Y., Truskolaski, A. & Boivin, P. (1985) Precipitation of plasminogen by Procion Red HE-3B. *J. Chromatogr.* **346**, 111–124

26. Pearson, J.C., Clonis, Y.D. & Lowe, C.R. (1989) Preparative affinity preparation of l-lactate dehydrogenase. *J. Biotechnol.* **11**, 267–274

27. Clonis, Y.D. & Lowe, C.R. (1980) Triazine dyes, a new class of affinity labels for nucleotide-dependent enzymes. *Biochem. J.* **191**, 247–251

28. Van Dam, M.E., Wuenschell, G.E. & Arnold, F.H. (1989) Metal affinity precipitation of proteins. *Biotechnol. Appl. Biochem.* **11**, 492–502

29. Lilius, G., Persson, M., Bülow, L. & Mosbach, K. (1991) Metal affinity precipitation of proteins carrying genetically attached polyhistidine affinity tails. *Eur. J. Biochem.* **198**, 499–504

30. Schneider, M., Guillot, C. & Lamy, B. (1981) The affinity precipitation technique. Application to the isolation and purification of trypsin from bovine pancreas. *Ann. N.Y. Acad. Sci. U.S.A.* **369**, 257–263

31. Senstad, C. & Mattiasson, B. (1989) Affinity-precipitation using chitosan as ligand carrier. *Biotechnol. Bioeng.* **33**, 216–220

32. Senstad, C. & Mattiasson, B. (1989) Purification of wheat germ agglutinin using affinity flocculation with chitosan and a subsequent centrifugation or flotation step. *Biotechnol. Bioeng.* **34**, 387–393

33. Taniguchi, M., Kobayashi, M., Natsui, K. & Fujii, M. (1989) Purification of staphylococcal protein A by affinity precipitation using a reversibly soluble–insoluble polymer with human IgG as a ligand. *J. Ferment. Bioeng.* **68**, 32–36

34. Nguyen, A.L. & Luong, J.H.T. (1989) Syntheses and applications of water-soluble reactive polymers for purification and immobilisation of biomolecules. *Biotechnol. Bioeng.* **34**, 1186–1190

35. Chen, J.P. & Hoffman, A.S. (1990) Polymer–protein conjugates. II. Affinity precipitation separation of human immunogammaglobulin by a poly (*N*-isopropylacrylamide)–protein A conjugate. *Biomaterials* **11**, 631–634

36. Monji, N. & Hoffman, A.S. (1987) A novel immunoassay system and bioseparation process based on thermal phase separating polymers. *Appl. Biochem. Biotechnol.* **14**, 107–120

37. Senstad, C. & Mattiasson, B. (1989) Preparation of soluble affinity complexes by a second affinity interaction: a model study. *Biotechnol. Appl. Biochem.* **11**, 41–48

38. Bradshaw, A.P. & Sturgeon, R.J. (1990) The synthesis of soluble polymer–ligand complexes for affinity precipitation studies. *Biotechnol. Tech.* **4**, 67–71

39. Linné, E., Garg, N., Kaul, R. & Mattiasson, B. (1992) Evaluation of alginate as a ligand carrier in affinity precipitation. *Biotechnol. Appl. Biochem.* **16**, 48–56

40. Gupta, M.N., Dong, G.Q. & Mattiasson, B. (1993) Purification of endo-polygalacturonase by affinity precipitation using alginate. *Biotechnol. Appl. Biochem.* **18**, 321–328

41. Kamihira, M., Kaul, R. & Mattiasson, B. (1992) Purification of recombinant protein A by aqueous two-phase extraction integrated with affinity precipitation. *Biotechnol. Bioeng.* **40**, 1381–1387

9

Molecular pathology of prion diseases

*Corinne Smith and †John Collinge

Department of Biochemistry, School of Medical Sciences, University of Bristol, Bristol BS8 1TD, U.K. and †Department of Biochemistry and Molecular Genetics, St Mary's Hospital Medical School, Imperial College, London W2 1PG, U.K.

Introduction

In 1959 William Hadlow drew a connection between a degenerative neurological disorder in sheep and goats called scrapie and a human disease, kuru, affecting the Fore peoples of Papua New Guinea. Scrapie had been shown to be experimentally transmissible, by inoculation, between sheep and goats in 1936 and this led to the suggestion that kuru might also be transmissible. Kuru was successfully transmitted by intracerebral inoculation to chimpanzees by Gajdusek and colleagues in 1966; Creutzfeldt–Jakob disease (CJD), a rare cause of human dementia which has similar histological features to scrapie, was transmitted in 1968. This remarkable work led to the development of the concept of the transmissible spongiform encephalopathies, also called the transmissible dementias or slow virus diseases. The term spongiform encephalopathy describes a common histological feature of the diseased brains in which vacuoles are seen — giving the brain a 'sponge-like' appearance.

A family of related animal diseases has now been recognized, including scrapie in sheep and goats and similar diseases in mink, mule deer, elk and cattle. The bovine form of the disease, bovine spongiform encephalopathy (BSE) or 'mad cow disease', was first observed in the United Kingdom in 1985 and the number of cases has risen steadily to produce a major epidemic that has now affected over 100000 cows. The cause of this epidemic was traced to a change in the rendering process used to produce meat and bone-meal from the carcasses of sheep, some of which would have been scrapie contaminated[1]. The meat and bone-meal were used as a dietary supplement for dairy cows and other animals. In particular, hydrocarbon solvent extraction was discontinued and this appears to have allowed more of the scrapie agent to enter the finished foodstuff. Prior to identification of the disease, BSE-affected cows also entered the rendering process and further fuelled the epidemic. Since material from affected cows undoubtedly entered the human food-chain before the introduction of appropriate restrictions, the possibility that human cases will result has highlighted research in this field. Recently reported spongiform encephalopathies of a number of zoo animals have, almost certainly, also arisen from this use of contaminated foodstuffs. Although the disease can be passed between mammalian species it does so with difficulty, following long incubation periods: the so-called 'species barrier'. The effectiveness of the bovine-to-human barrier remains to be established.

Gerstmann–Sträussler syndrome (GSS) was transmitted to experimental animals in 1981 and, with CJD and kuru, completes the trio of traditionally recognized human spongiform encephalopathies. Kuru is recognized by a progressive impairment of motor co-ordination, cerebellar ataxia, although dementia can occur later in some individuals. CJD is characterized by a rapidly progressive dementia and, in most cases, the clinical course is less than 6 months. GSS is a more protracted illness with a mean duration of 5 years, although some cases continue for more than 20 years. Classically, GSS is regarded as a chronic cerebellar ataxia with dementia occurring later in the course of the disease. The diseases are relentlessly progressive, invariably fatal and no known treatment affects the outcome.

Kuru was transmitted among the Fore peoples by ritualistic cannibalism. These practices probably involved both ingestion of tissues and individuals smearing themselves with tissues from deceased relatives in mourning rituals. Kuru reached epidemic proportions in certain villages, being the major cause of death of children and young adults, but has now become extremely rare since the cessation of cannibalistic practices in the late 1950s (for review, see reference 2). A few cases still occur each year, however, because of the extremely prolonged incubation period (which can exceed 40 years).

CJD usually occurs as a sporadic illness with no history of exposure to either a human or animal source of the infective agent. Epidemiological studies do not show the clustering which would be expected for an infectious disease: the incidence of sporadic CJD is remarkably uniform throughout the world[3]. In particular, there is no correlation between the incidence of CJD and local scrapie prevalence. However, around 15% of cases of CJD and nearly all cases of GSS occur in a familial context and show an autosomal dominant pattern of disease segregation.

The nature of the infective agent has been the subject of much debate. It was naturally assumed that the transmissible agent must be some sort of 'slow virus'. Against this interpretation, however, were both the failure to identify such a virus and the remarkable properties of the infective agent. In particular, infectivity was remarkably resistant to treatments that would inactivate the putative nucleic acid genome of such a virus (for example, u.v.-irradiation or treatment with nucleases), but sensitive to treatments that denature proteins. This led a number of workers to speculate in the 1960s that the agent may be devoid of nucleic acid and it was indeed suggested that the agent may be simply a protein. Clearly such notions were controversial, and arguments for some type of atypical virus were supported by the demonstration by Dickinson and others of the existence of several different 'strains' of experimental scrapie in mice. These isolates had characteristic and stable properties (in particular with respect to incubation period and pattern of histopathology) and led many to assume that these phenotypic properties must be encoded by an agent-specific nucleic acid (for recent review, see reference 4). Progressive enrichment of brain fractions for infectivity led to the identification of a protein by Prusiner and co-workers in 1982, now known as the prion protein (PrP). This accumulates in the brains of affected animals or humans, sometimes forms amyloid plaques and is partially resistant to proteolysis with proteinase K. At the time it was assumed that this protein was virally encoded, but partial N-terminal sequencing allowed the design of isocoding mixtures of oligonucleotides which were used to screen a cDNA library from scrapie-infected hamsters. Remarkably, PrP was shown to be encoded by a single copy, host chromosomal gene. Protease-resistant PrP had an apparent molecular mass on SDS/PAGE of 27–30 kDa and became known as PrP^{27-30}; however, the normal cellular isoform of the protein, PrP^C, was found to be fully protease sensitive and to have a molecular mass of 33–35 kDa. A number of experiments showed that the abnormal, disease-related isoform, designated PrP^{Sc}, was the principal component of the infective fractions[5]. PrP^{Sc} was also of molecular mass 33–35 kDa, and PrP^{27-30} was derived from it by limited proteolysis. Prusiner introduced the term 'prion' in 1982 to distinguish these *proteinaceous infec*tious particles from both viruses and viroids. To some degree the debate continues but, since 1982, evidence has mounted in favour of the hypothesis that the agent which causes scrapie is an abnormal isoform of a host-encoded protein.

Molecular genetics of the human prion diseases

After cloning the hamster PrP gene in 1985, the human PrP cDNA was cloned and sequenced in 1986. Since some CJD cases are familial and GSS is usually familial — both demonstrate an autosomal dominant pattern of disease segregation — it was reasonable to perform linkage studies for an autosomal gene that may cause these disorders. The PrP gene was an obvious candidate gene and, in 1989, genetic linkage was demonstrated between GSS and a mis-sense mutation at codon 102 (see Figure 1) of the PrP gene in two kindreds[6]. An insertional mutation in PrP had previously been reported in a British family with CJD[7]. Both the insertion and the codon 102 mutation (Pro-102 → Leu)

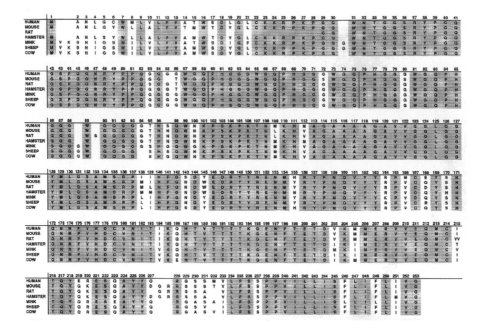

Figure 1. Alignment of human, mouse, rat, Syrian hamster, mink, sheep and cow PrP amino acid sequences
The shaded areas indicate residues which are the same at that position in all species. The numbering applies to the human sequence. The N-terminal repeat region commences at around residue 53 and continues to residue 93.

were not present in normal control populations. Therefore, it seemed likely that they represented pathogenic mutations. Direct demonstration of the pathogenicity of the Leu-102 PrP variant came with the production of transgenic mice with PrP transgenes encoding leucine at the corresponding murine codon. These mice spontaneously developed spongiform neurodegeneration at around 160 days and PrP[Sc] was detected, albeit at low levels, in their brains[8]. Furthermore, it has now been demonstrated that brain tissue from such transgenic mice can transmit the disease to wild-type animals by intracerebral inoculation[9]. Hence the infectious agent had been effectively synthesized by expressing mutated PrP in mice, providing compelling evidence for the proposition that infectivity is due to PrP[Sc].

Inherited prion diseases
The discovery of the first pathogenic PrP mutations was rapidly followed by the elucidation of a large number of both insertional and point mutations in inherited cases of prion disease[10]. The insertions are of extra copies of octapeptide repeat elements found in an array of five tandem repeats in the normal protein (see Figures 1 and 2). Insertions discovered so far include individuals with between two and nine extra repeats. Point mutations result in amino-acid

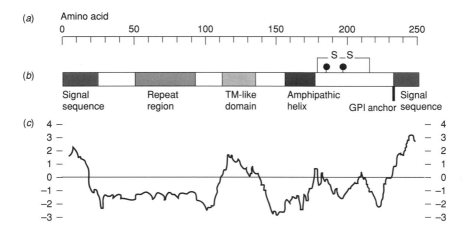

Figure 2. Diagram of domain patterns within the primary sequence of PrP (based on studies on Syrian hamster PrP)

(a) The sequence commences with a 22 amino acid signal peptide which is removed during biosynthesis. This is followed by a region containing five tandem octapeptide repeats. There follows a highly hydrophobic section which has been predicted to be a transmembrane helix and a section predicted to form an amphipathic helix. (b) The protein contains a disulphide bridge (S–S) and asparagine-linked glycosylation occurs at two sites (●). A hydrophobic signal sequence at the C-terminal end of the protein is removed on addition of the GPI anchor at position 231. (c) A hydropathy plot after the method of Kyte and Doolittle. This provides a measure of the hydrophobic character of each section of the sequence.

substitutions (or, in one case, in a stop codon, with production of a truncated protein) in PrP at residues 102, 105, 117, 145, 178, 180, 198, 200, 210, 217 and 232 (see Table 1).

The availability of direct gene markers for the inherited diseases has enabled study of the phenotypic range of these disorders, and it has been possible to demonstrate remarkable heterogeneity within families and also to identify some previously unrecognized prion diseases. Initial observations in this regard were made in a kindred with a 144 bp insertion in the PrP gene[7]. While a case in this kindred had classical features of CJD, another affected individual from this family had an illness that lasted for over 4 years but, at autopsy, only mild and subtle spongiform changes (insufficient for a histological diagnosis of CJD) were seen. An additional family member has had the illness for over 12 years. Therefore, a range of illnesses — from classical subacute CJD to a GSS-like disorder (indeed the example given exceeds the longest previously published clinical duration for GSS) — can co-exist in the same family with the same genetic mutation. Additionally, the histological features can vary from gross to minimal. For this reason it was argued that other GSS/CJD-type illnesses might present atypically and mimic other neurodegenerative conditions. This led to screening of neurodegenerative cases and the identification of families with mutations in the PrP gene who were not thought, on clinical grounds, to be suffering from spongiform encephalopathies. The first family to be detected in this way using PrP gene analysis had

Table 1. Point mutations in the PrP gene which are known to cause human prion disease

Position	Amino acid change	
	Native	Mutant
102	Proline	Leucine
105	Proline	Leucine
117	Alanine	Valine
145	Tyrosine	Stop
178	Aspartate	Asparagine
180	Valine	Isoleucine
198	Phenylalanine	Serine
200	Glutamate	Lysine
210	Valine	Isoleucine
217	Glutamine	Arginine
232	Methionine	Arginine

been thought clinically to be suffering from familial Alzheimer's disease[11]. Further screening of over 100 cases revealed another four families with an identical PrP gene insertion, and subsequent genealogical investigation demonstrated that all of these cases formed part of a single large kindred in the South-East of England[12]. The demonstration that an individual from this family lacked any histological features of CJD, GSS or any other specific neurodegenerative condition raised the possibility that prion diseases were under-diagnosed, as perhaps 10% of patients dying with dementia who undergo autopsy do not have sufficient histological findings to reach a diagnosis[13]. Therefore, diagnostic terms such as CJD, GSS and spongiform encephalopathy are becoming less useful with the demonstration that CJD and GSS form part of a wider spectrum of disease and can co-exist in the same family with the same mutation. The histological features regarded as being characteristic of these disorders are not invariably present. Since an aberrant form of the PrP and its gene play a key role in the aetiology of these conditions, prion disease, which can then be subdivided into acquired, sporadic and inherited forms, seems a more suitable terminology.

While cases within the 144 bp insertion pedigree clinically resembled Alzheimer's disease, a number of types of inherited prion disease have now been identified that histologically resemble Alzheimer's disease in many other respects. For instance, inherited prion diseases with the codon 198 or codon 217 mutations are associated with the presence of neurofibrillary tangles, indistinguishable morphologically or antigenically from those seen in Alzheimer's disease.

Of particular interest has been the finding of a mutation at codon 178 in several families with what is described as fatal familial insomnia[14]. This disease is characterized by progressive untreatable insomnia and autonomic dysfunction. It has now been reclassified as one of the inherited prion diseases and, interestingly, carries the same mutation as that seen in several families from Northern Europe with a phenotype similar to CJD[15]. Such findings emphasize the need for a molecular re-classification of neurodegenerative diseases based on aetiological rather than descriptive criteria.

While CJD usually occurs as a sporadic disease without clustering of cases, there are three well-documented ethnogeographic clusters of CJD that are an apparent exception to this random distribution of cases: among Libyan Jews, in an area of Slovakia and in Chile. At least in the case of the Libyan Jews this high incidence was thought to be due to dietary habits, including eating sheep eyeballs and lightly cooked sheep brain. All three clusters are now known to be familial and associated with the codon 200 mutation.

The availability of such direct gene tests for the inherited forms of these diseases allows not only unequivocal diagnosis but also presymptomatic testing of unaffected but at risk family members[16]; in addition, it opens up the potential for antenatal testing. In some families it may also be possible to determine whether a gene carrier will have an early or late onset of disease by codon 129 genotyping (discussed later).

As well as pathogenic PrP mutations, a number of coding and non-coding polymorphisms have been described. A common protein polymorphism at residue 129 is known to be important in genetic susceptibility to prion diseases (see below). Deletions of a single octapeptide repeat element have been reported in both patients and normal controls and represent uncommon polymorphisms present in around 1% of normal Caucasians[17].

Iatrogenic prion diseases

The remarkable resistance of prions to standard sterilization methods has led to a number of accidental cases of prion disease resulting from medical treatment. More than 42 cases have arisen through injections of human cadaveric pituitary-derived growth hormone, some batches of which were contaminated with the CJD agent. Two cases have been reported following treatment with human pituitary-derived gonadotrophin as treatment for infertility. Other iatrogenic routes include dura mater and corneal grafting and the use of inadequately sterilized neurosurgical instruments.

In the U.K. there have been 10 cases of growth-hormone-related CJD, out of a total of 1908 people treated before the risk of transmitting CJD by this route was discovered in 1985. Each batch of growth hormone was pooled from up to 3000 pituitary glands removed at routine autopsy. The individuals affected had received hormone from more than one batch and no batch spanned all cases. This implies that all patients treated with hormone derived in this way may have been exposed to the CJD agent. It is possible that the recipients who developed CJD received a greater dose but it is also possible that they were genetically predisposed to succumb to infection. To investigate this, the six cases known in the U.K. at the time of the study, and a case that had resulted from treatment with pituitary-derived gonadotrophin, were screened for variations in the PrP gene. No known pathogenic mutation was found in these samples, but variation in a coding polymorphism at codon 129 proved of interest. In the Caucasian population there is a genotype frequency at this codon of 37% Met/Met, 51% Met/Val and 12% Val/Val (estimated from 106 controls). However, four of the seven iatrogenic CJD cases analysed had the Val/Val genotype, while two were Met/Val heterozygotes and one was a Met/Met homozygote[18].

Although the numbers are small this suggested that Val/Val homozygotes may be genetically predisposed to contract prion disease on exposure to the infectious agent. These results have since been confirmed by other workers[19]. It will be of interest to see whether kuru occurred principally in individuals with this genotype.

Sporadic prion diseases

This term describes all prion diseases to which no inherited or clear infectious origin can be attributed and their aetiology is unclear. However, as with iatrogenic CJD there is evidence for genetic susceptibility. Analysis of the genotype at codon 129 of 22 sporadic CJD cases revealed that only one was heterozygous with respect to this common protein polymorphism, compared with 51% heterozygosity in the normal population[20]. This suggests that homozygotes for this polymorphism may be genetically susceptible to sporadic prion disease, while heterozygotes are in some way protected.

In addition to establishing that individuals homozygous at codon 129 may be susceptible to developing prion disease, the genetic information gained by analysing sporadic and iatrogenic cases has a wider significance concerning the mechanism of prion propagation; there is strong support for a model of prion propagation based on the importance of both a direct protein–protein interaction and primary structure identity for the efficiency of this interaction (discussed in detail below).

It remains a possibility that some cases arise from environmental prion exposure, for instance from a dietary source, although epidemiological studies provide no evidence to support this. Alternative hypotheses proposed for the aetiology of sporadic CJD include spontaneous conversion of PrP^C to PrP^{Sc} as a rare stochastic event or somatic mutation of the PrP gene; and such explanations would be more consistent with the apparently random case distribution seen in epidemiological studies. The recent demonstration that mice overexpressing normal hamster PrP develop a novel prion disease[21] raises the possibility that some cases of sporadic CJD arise from overexpression of wild-type PrP^C.

A model of prion propagation

Some of the most informative work on prion disease has come from the use of transgenic mice expressing foreign PrP genes to manipulate the species barrier to transmission.

Transmission of prion diseases between mammalian species is possible, but difficult. Typically only a small number of inoculated animals become sick, and then only after prolonged incubation periods. This resistance to development of disease on initial challenge is referred to as the species barrier. When material from such an infected animal is used to inoculate another animal of the same species, however, disease nearly always results, with a shorter incubation time, which then usually remains constant for subsequent passages.

Prusiner and colleagues constructed transgenic mice containing Syrian hamster PrP transgenes in addition to the normal mouse PrP gene. These mice were inoculated with either mouse or Syrian hamster prions and monitored for development of disease (see Figure 3). The normal control mice were unaffected by the hamster inoculation but all died within 150 days of inoculation with mouse prions. The mice expressing Syrian hamster PrP were still susceptible to

Figure 3. Overcoming the species barrier through the use of transgenic mice
(a) Mice inoculated with mouse prions contract disease but are not affected by hamster prions. (b) The reverse is true for hamsters which are susceptible to hamster prions but not to mouse prions. (c) Transgenic mice expressing hamster PrP as well as mouse PrP are susceptible to prions from either species. Prions isolated from the brains of diseased mice reflect the species used for the inoculum.

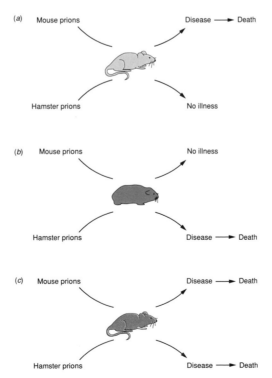

mouse prions but now became fully susceptible to infection by hamster prions — with incubation periods inversely proportional to transgene copy number. Furthermore, infectivity produced from such transgenic mice was determined by the inoculum: transgenic mice inoculated with mouse prions produced infectivity fully pathogenic for wild-type mice but not for hamsters, while inoculation with hamster prions resulted in infectivity now fully pathogenic for hamsters but not mice. Thus the species barrier was removed by expression of PrP identical in sequence to that of the inoculum. Such observations, allied with the importance of homozygosity with respect to a common PrP polymorphism in both the initiation and progression of the human prion diseases, has provided support for a model of prion propagation based on direct PrP–PrP interaction (Figure 4). Sporadic prion disease occurs principally in individuals homozygous with respect to a common polymorphism at residue 129. Furthermore, in the type of inherited prion disease associated with a 144 bp insertional mutation, onset of disease is 10–20 years later for individuals heterozygous with respect to this protein polymorphism than for homozygotes[12,22]. It is proposed that PrP^{Sc}, whether introduced by inoculation or formed spontaneously from mutant PrP, can act to induce the conversion of further PrP^{C} molecules to PrP^{Sc} leading to their progressive accumulation. PrP^{Sc} is known to be derived from PrP^{C} by a post-translational process and the precise nature of this change is yet to be elucidated.

Effect of disruption of PrP gene in mice

In 1992 mice were generated in which the host PrP gene was inactivated by replacement of about two-thirds of the PrP coding sequence with a neomycin-resistance gene, using homologous recombination techniques in embryonic stem cells. Since PrP is a highly conserved protein among mammals, and is known to be expressed at an early stage in embryogenesis, it was assumed that homozygosity for such disrupted PrP genes would give a severe, perhaps embryonic lethal phenotype. Surprisingly, homozygous PrP gene-ablated mice were not only viable but appeared essentially normal phenotypically[23]. Explanations proposed include that the protein is now obsolete and that its conservation results from evolutionary inertia; the opposite argument is that its function is so crucial that a high degree of redundancy exists such that other proteins can take over its function. The absence of any clear phenotype for the mice makes these questions hard to answer at present and the function of PrP^{C} remains enigmatic.

The normality and fertility of such PrP-ablated mice does, however, allow a direct test of the idea that PrP is essential for infectivity. The results of such experiments have now been reported and, indeed, such mice neither replicate the agent after inoculation nor develop prion disease. While this result is entirely consistent with the 'protein only' hypothesis of prion replication, it could be argued that PrP constitutes a viral receptor and, therefore, that its deletion would equally protect against infection with the putative scrapie virus.

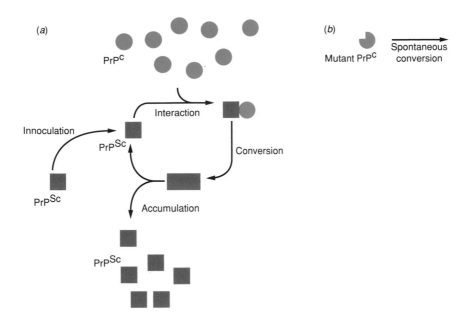

Figure 4. Diagram to explain the protein only model of prion replication

(a) PrP^C (represented by blue circles) binds to PrP^{Sc} (represented by grey squares) which in this case has been introduced into the organism by inoculation. There follows the imposition on PrP^C of the structure of PrP^{Sc} as a result of this interaction but by a mechanism not yet understood. This is succeeded by a chain reaction of PrP^{Sc} accumulation as the newly converted PrP^{Sc} molecules bind to further PrP^C molecules. (b) In the case of inherited prion disease, a mutation in the PrP gene results in a predisposition for the mutant form of PrP^C to convert spontaneously to PrP^{Sc}. The chain reaction shown in (a) ensues and PrP^{Sc} accumulates.

Thus experiments do not discriminate between the protein only and virus models. A further important result from this experiment was that mice with a single copy of the ablated gene (and which had approximately 50% of normal PrP expression) had both extremely prolonged incubation periods and long duration of clinical illness. This implies that these mice are in some way protected from infection as a result of the decreased level of PrP^C expression — further indication of the importance of protein–protein interactions in the propagation of the disease and that the process of prion generation relies on recruitment of normal protein by PrP^{Sc}.

Analysis of prion structural elements

If a highly specific protein–protein interaction is part of the process which triggers prion disease, knowledge of the precise 3-dimensional structure of the infectious agent is crucial to tackling the disease. This unfortunately remains elusive to date, since most techniques which provide information on protein structure require high concentrations of soluble protein. PrP^{Sc} is extremely insoluble and has a strong tendency to aggregate. Purification has led to the production of PrP$^{27-30}$, which is a truncated version of the full protein but which retains infectivity. This truncation is a result of the limited proteolysis used to purify the disease-related protein. The majority of biophysical studies have been carried out using this version of the protein.

Of fundamental importance is the question of whether isolated PrP^{Sc} is identical in composition to the translated gene sequence. To this end Stahl and co-workers[24] carried out detailed analysis of purified PrP^{Sc} and PrP$^{27-30}$ from Syrian golden hamster using mass spectrometry and amino acid sequencing. They found that the primary sequence was identical to that expected from the translated gene sequence and could identify no post-translational chemical modifications that distinguished PrP^{Sc} from PrP$^{C.}$ These results suggest that the differences between the infectious and normal cellular protein lie in the 3-dimensional conformation. The transition of the normal protein to its pathogenic form may, therefore, be due to a rearrangement of the protein fold.

Primary structure

Figures 1 and 2 show the primary sequence and some structural features of hamster PrP. The sequence of PrP^{C} is known for a number of species and there is greater than 80% sequence identity between mammals; the chicken sequence, however, shows only around 30% identity. Although the protein is highly conserved across mammalian species it is not related by sequence to any other known protein. PrP is glycosylated at Asn-159 and Asn-173 and has a glycosylphosphatidylinositol (GPI) anchor attached (at Ser-231 in the hamster sequence).

In the absence of a 3-dimensional structure, it is of interest to attempt to predict the nature of structural domains from the protein sequence. The limitations of methods for 2-dimensional structural prediction must be noted; in particular, α-helices tend to be over-predicted. However, they work well enough to provide a focus for discussion and experiment. Bazan and colleagues[25] used a range of computational analyses to assess the potential secondary structure of hamster PrP. They predicted two transmembrane domains on the basis of these analyses. One domain (residues 90–103) was predicted to form a hydrophobic transmembrane helix, the other (residues 119–147) was predicted to form an amphipathic α-helix across the membrane. However, the majority of PrP^{C} molecules can be released from the cell membrane by phospholipase C which cleaves at the phospholipid anchor[26]. Hence, at most, only a very small proportion of PrP molecules could be incorporated into the membrane. This is surprising given the confidence with which the transmembrane domain could be predicted.

In addition to the more easily predicted sequence patterns, the N-terminal section of PrP contains a series of five octapeptide repeats which have no significant similarity to known protein sequences. No function for this unique motif has been proposed. PrP$^{27-30}$, in which most or all of this region is removed, retains infectivity. However, insertions of integral numbers of these repeats in tandem cause disease. Also of considerable interest is the discovery of a stop codon at position 145 and the production of a protein truncated at the C-terminus in cases of inherited prion disease[27].

Secondary structure

The secondary structure content of both the cellular and infectious isoforms of the protein has been estimated by circular dichroism (c.d.) spectroscopy and attenuated total reflection Fourier transform infrared spectroscopy (a.t.r.-f.t.i.r.), respectively. The c.d. spectrum of PrP[C] showed this protein to be predominantly α-helical (estimated to be 42%) with little β-sheet present (estimated to be 3%)[28]. In contrast Gasset and co-workers[29] showed that PrP[27–30] (the protease-resistant core of PrP[Sc] which is normally purified as a result of a limited proteolysis step) contained 54% β-sheet, 25% α-helix, 10% turns and 11% random coil. A previous f.t.i.r. study[30] in which a suspension of prion rods in phosphate-buffered saline was used, produced estimates of 48% β-sheet, 14% α-helix, 34% turn and 4% random coil. To further understand the processes by which disease is caused, the a.t.r.-f.t.i.r. samples of PrP[27–30] were subjected to a range of conditions known to affect infectivity, namely high pH (up to pH 12) and denaturation with SDS. The effect of dispersal into lipid complexes was also studied. The β-structure produces two types of signal: that due to intermolecular aggregation (which gives a low-frequency signal) and that due to β-sheet structure within the protein molecule. About 60% of the total β-sheet content of PrP[27–30] is due to this low-frequency, aggregated β-sheet. In general it was found that conditions which reduced this type of β-structure also resulted in disassembly of prion rods and loss of infectivity. Correlation of changes in secondary structure with changes in infectivity, together with the altered conformation of PrP[27–30], give weight to the proposal that the protein's conformation must be of central importance to its pathogenicity.

Expression of PrP domains

A more biologically relevant way of investigating the role of primary structure in the synthesis of PrP[Sc] is that described by Rogers and colleagues[31] in which a form of PrP is expressed in a tissue culture system (mouse neuroblastoma cells) infected with exogenous prions. They monitored the production of protease-resistant PrP on expression of truncated versions of the PrP gene. These included removal of the N-terminal 66 amino acids (the octapeptide repeat region) and removal of the GPI anchor attachment signal (which prevents attachment of the GPI anchor). None of these modifications prevented formation of protease-resistant PrP. Furthermore, one, two, four and six additional octapeptide repeat units were introduced into the construct to investigate why such insertions precipitate disease. The efficiency of production of PrP[Sc] was, however, not altered by the addition of octapeptide repeat units. The conclusion is that the regions of the PrP gene which are required for production of protease-resistant protein lie in the central two-thirds of the sequence and do not include the octapeptide repeat region or the GPI anchor. It should also be noted that these regions are not required for infectivity of the protein, since they are removed on production of PrP[27-30]. The fact that mutations affecting these regions cause disease must, however, still be explained. In this regard it should be noted that the tissue culture was already infected with prions. It may be that such mutations are instrumental in the initial spontaneous transition from normal cellular PrP to the pathogenic form.

Summary of theories of prion propagation

Protein only hypothesis

Normal PrP undergoes a conformational rearrangement which renders the molecule infectious. In this case the term 'infectious' describes a process by which the transformed molecule binds tightly to one or more normal PrP molecules facilitating a similar structural change in the normal PrP. When this species comes into contact with other normal PrP molecules these too are converted into the tight-binding form of the protein. In this way a polymer forms which cannot be broken down by the cell, and infectious protease-resistant material gradually accumulates (see Figure 4). Nearly all the work described in this article points to a highly specific protein–protein interaction underlying the cause and propagation of this disease. This is underlined by the genetic susceptibility of homozygous individuals to iatrogenic and sporadic disease and the studies on the species barrier in transgenic mice. The fact that the protein aggregates into β-structure in the pathogenic form and that short fragments show a high tendency to aggregate also supports this idea.

Virus theory

Although nearly all the evidence points to a protein as the major factor in dictating the occurrence and progress of prion disease, it is possible to explain these data in terms of a slow virus with uncharacteristic transmission properties. The protein may be required as a receptor for the virus or contain some DNA which is completely protected by the protein from nucleases or u.v.-irradiation. The existence of different strains of PrP[Sc] is more easily explained by a virus theory. However, extensive searches for any DNA within scrapie protein[32] has shown that for one DNA molecule to be present per infective unit the length of such a fragment could not be greater than 100 bp. It is increasingly difficult to propose that a virus or any significant agent-specific nucleic acid is involved, in the light of recent advances in both the human molecular genetics and murine transgenetics of these diseases, without at the same time presenting direct evidence for their existence.

The 'unified' theory

The unified theory of prion replication has been proposed by Weissmann[33] and is an adaptation of the protein only model. It is proposed that a second molecule or 'co-prion' is associated with the protein which can modify its behaviour and so account for the different phenotypic properties associated with the different strains or isolates of agent in experimental murine scrapie. The hypothetical co-prion is not required for infectivity but, when present in the inoculum, carries agent-specific information. If the co-prion is absent from the inoculum the 'apoprion' can recruit a co-prion from the host, and so takes on specific characteristics. The most likely candidate molecule would be a nucleic acid but, again, there is no direct evidence for the existence of such co-factors. However, this model leads to a key testable prediction, namely that while infectivity of particular prion isolates may not be sensitive to irradiation sufficient to destroy nucleic acids, their strain properties would be.

Conclusions

There is strong evidence that prion diseases are protein-based and that infectious particles contain no essential polynucleotide material. The initially puzzling fact — that the causative agent reproduces itself — can be explained by the concept that the pathogenic form of the protein imposes its structure on the cellular PrP provided by the host organism. This can explain how iatrogenic, genetic and sporadic diseases can be 'infectious' without the requirement for an agent-specific nucleic acid code.

The issue of the existence of different strains of experimental scrapie in mice has yet to be adequately explained; many workers therefore argue that there must be an agent-specific genome. However, modifications of PrP, such as different glycosylation patterns, which may be produced in specific neuronal populations, are under intensive investigation. It has been hypothesized that PrP^{Sc} produced by a specific neuronal population would preferentially target that same population in a subsequent inoculated host. Such a model would be consistent with the remarkable bilateral symmetry of pathological changes often seen following unilateral intracerebral inoculation. It is not known at present whether different strains of human prions are produced or whether they may be clinically relevant.

The key difference between normal cellular PrP^{C} and the infectious form of the protein resides in a post-translational modification, the nature of which has still to be defined. By a process of exclusion and by preliminary biophysical studies, their 3-dimensional structure or folding pattern seems the most likely candidate. Hence the conversion of PrP^{C} to PrP^{Sc} may involve a major conformational rearrangement or folding transition. The processes which cause this transition must be discerned in order to understand, and eventually prevent, these diseases. In short, prion diseases may be the first example of a transmissible 'protein folding' disease.

An interesting biological parallel is that of the p53 tumour suppressor protein. Mutations in p53 are associated with a wide range of tumours. It has been shown that some forms of mutant p53 adopt abnormal, and biologically inactive, conformations. Milner and Medcalf showed that mutant p53 can drive co-translated, wild-type p53 into the mutant conformation[34]. This provides a close parallel with the mechanism proposed for prion propagation and raises the possibility that the pathogenic mechanism proposed may not be unique to these rare neurodegenerative diseases.

For many people, final demonstration of the 'protein only' theory of prion replication will be the ability to generate prions synthetically by protein engineering. If PrP could be generated in an expression system and shown to be infectious there would be no further doubt concerning the role of nucleic acid in this process. Such a system would subsequently allow a comprehensive analysis of the regions of the protein required for both the ability to convert normal PrP to the pathogenic form and the initial transition to the infectious form. This will pave the way for effective therapies to be developed for the prion diseases.

Further reading

Prusiner, S.B, Collinge, J., Powell, J. & Anderton, B. (eds.), *Prion Diseases of Humans and Animals* (1992), Ellis Horwood, London

Collinge, J. & Weissmann, C. (1994) Molecular biology of prion diseases. *Phil. Trans. R. Soc. London* B **343**, 357–463

References

1. Wilesmith, J.W., Ryan, J.B. & Atkinson, M.J. (1991) Bovine spongiform encephalopathy: epidemiological studies on the origin. *Vet. Rec.* **128**, 199–203
2. Alpers, M.P. (1992) Reflections and highlights: a life with kuru. In *Prion Diseases of Humans and Animals* (Prusiner, S.B., Collinge, J., Powell, J. & Anderton, B., eds.), Ellis Horwood, London
3. Brown, P., Cathala, F., Raubertas, R.F., Gajdusek, D.C. & Castaigne, P. (1987) The epidemiology of Creutzfeldt-Jakob disease: conclusion of a 15-year investigation in France and review of the world literature. *Neurology* **37**, 895–904
4. Bruce, M.E. & Fraser, H. (1991) Scrapie strain variation and its implications. *Curr. Top. Microbiol. Immunol.* **172**, 125–138
5. Prusiner, S.B. (1987) Prions and neurodegenerative diseases. *N. Engl. J Med.* **317**, 1571–1581
6. Hsiao, K., Baker, H.F., Crow, T.J., Poulter, M., Owen, F., Terwilliger, J.D., Westaway, D., Ott, J. & Prusiner, S.B. (1989) Linkage of a prion protein missense variant to Gerstmann–Straussler syndrome. *Nature (London)* **338**, 342–345
7. Owen, F., Poulter, M., Lofthouse, R., Collinge, J., Crow, T.J., Risby, D., Baker, H.F., Ridley, R.M., Hsiao, K. & Prusiner, S.B. (1989) Insertion in prion protein gene in familial Creutzfeldt–Jakob disease. *Lancet* **i**, 51–52
8. Hsiao, K.K., Scott, M., Foster, D., Groth, D.F., DeArmond, S.J. & Prusiner, S.B. (1990) Spontaneous neurodegeneration in transgenic mice with mutant prion protein. *Science* **250**, 1587–1590
9. Hsiao, K.K., Groth, D., Scott, M., Yang, S.L., Serban, A., Rapp, D., Foster, D., Torchia, M., DeArmond, S.J. & Prusiner, S.B.(1992) Genetics and transgenic studies of prion proteins in

Gerstmann–Sträussler–Scheinker disease. In *Prion Diseases of Humans and Animals* (Prusiner, S.B., Collinge, J., Powell, J. & Anderton, B., eds.), pp. 120–128, Ellis Horwood, London

10. Collinge, J. & Palmer, M.S. (1992) Prion diseases. *Curr. Opin. Genet. Dev.* **2**, 448–453

11. Collinge, J., Harding, A.E., Owen, F., Poulter, M., Lofthouse, R., Boughey, A.M., Shah, T. & Crow, T.J. (1989) Diagnosis of Gerstmann–Straussler syndrome in familial dementia with prion protein gene analysis. *Lancet* **ii**, 15–17

12. Poulter, M., Baker, H.F., Frith, C.D., Leach, M., Lofthouse, R., Ridley, R.M., Shah, T., Owen, F., Collinge, J., Brown, J., *et al.* (1992) Inherited prion disease with 144 base pair gene insertion: genealogical and molecular studies. *Brain* **115**, 675–685

13. Collinge, J., Owen, F., Poulter, M., Leach, M., Crow, T.J., Rossor, M.N., Hardy, J., Mullan, M.J., Janota, I. & Lantos, P.L. (1990) Prion dementia without characteristic pathology. *Lancet* **336**, 7–9

14. Medori, R., Montagna, P., Tritschler, H.J., LeBlanc, A., Cortelli, P., Tinuper, P., Lugaresi, E. & Gambetti, P. (1992) Fatal familial insomnia: a second kindred with mutation of prion protein gene at codon 178. *Neurology* **42**, 669–670

15. Goldfarb, L.G., Haltia, M., Brown, P., Nieto, A., Kovanen, J., McCombie, W.R., Trapp, S. & Gajdusek, D.C. (1991) New mutation in scrapie amyloid precursor gene (at codon 178) in Finnish Creutzfeldt–Jakob kindred. *Lancet* **337**, 425

16. Collinge, J., Poulter, M., Davis, M.B., Baraitser, M., Owen, F., Crow, T.J. & Harding, A.E. (1991) Presymptomatic detection or exclusion of prion protein gene defects in families with inherited prion diseases. *Am. J Hum. Genet.* **49**, 1351–1354

17. Palmer, M.S., Mahal, S.P., Campbell, T.A., Hill, A.F., Sidle, K.C.L., Laplanche, J.L. & Collinge, J. (1993) Deletions in the prion protein gene are not associated with CJD. *Hum. Mol. Genet.* **2**, 541–544

18. Collinge, J., Palmer, M.S. & Dryden, A.J. (1991) Genetic predisposition to iatrogenic Creutzfeldt–Jakob disease. *Lancet* **337**, 1441–1442

19. Brown, P., Preece, M.A. & Will, R.G. (1992) 'Friendly fire' in medicine: hormones, homografts, and Creutzfeldt–Jakob disease. *Lancet* **340**, 24–27

20. Palmer, M.S., Dryden, A.J., Hughes, J.T. & Collinge, J. (1991) Homozygous prion protein genotype predisposes to sporadic Creutzfeldt–Jakob disease. *Nature (London)* **352**, 340–342

21. Westaway, D., DeArmond, S.J., Cayetano-Canlas, J., Groth, D., Foster, D., Yang, S., Torchia, M., Carlson, G.A. & Prusiner, S.B. (1994) Degeneration of skeletal muscle, peripheral nerves and the central nervous system in transgenic mice overexpressing wild-type prion proteins. *Cell* **76**, 117–129

22. Collinge, J., Brown, J., Hardy, J., Mullan, M., Rossor, M.N., Baker, H., Crow, T.J., Lofthouse, R., Poulter, M., Ridley, R., *et al.* (1992) Inherited prion disease with 144 base pair gene insertion: clinical and pathological features. *Brain* **115**, 687–710

23. Bueler, H., Fischer, M., Lang, Y., Bluethmann, H., Lipp, H., DeArmond, S.J., Prusiner, S.B., Aguet, M. & Weissmann, C. (1992) Normal development and behaviour of mice lacking the neuronal cell-surface PrP protein. *Nature (London)* **356**, 577–582

24. Stahl, N., Baldwin, M.A., Teplow, D.B., Hood, L., Gibson, B.W., Burlingame, A.L. & Prusiner, S.B. (1993) Structural studies of the scrapie prion protein using mass spectrometry and amino acid sequencing. *Biochemistry* **32**, 1991–2002

25. Bazan, J.F., Fletterick, R.J., McKinley, M.P. & Prusiner, S.B. (1987) Predicted secondary structure and membrane topology of the scrapie prion protein. *Protein Eng.* **1**, 125–135

26. Caughey, B., Neary, K., Buller, R., Ernst, D., Perry, L.L., Chesebro, B. & Race, R.E. (1990) Normal and scrapie-associated forms of prion protein differ in their sensitivities to phospholipase and proteases in intact neuroblastoma cells. *J. Virol.* **64**, 1093–1101

27. Kitamoto, T., Iizuka, R. & Tateishi, J. (1993) An amber mutation of prion protein in Gerstmann–Straussler syndrome with mutant PrP plaques. *Biochem. Biophys. Res. Commun.* **192**, 525–531

28. Pan, K., Baldwin, M., Nguyen, J., Gasset, M., Serban, A., Groth, D., Mehlhorn, I., Huang, Z., Fletterick, R.J., Cohen, F.E. & Prusiner, S.B. (1993) Conversion of α-helices into β-sheets features in the formation of the scrapie prion proteins. *Proc. Natl. Acad. Sci. U.S.A.* **90**, 10962–10966

29. Gasset, M., Baldwin, M.A., Fletterick, R.J. & Prusiner, S.B. (1993) Perturbation of the secondary structure of the scrapie prion protein under conditions that alter infectivity. *Proc. Natl. Acad. Sci. U.S.A.* **90**, 1–5

30. Caughey, B.W., Dong, A., Bhat, K.S., Ernst, D., Hayes, S.F. & Caughey, W.S. (1991) Secondary structure analysis of the scrapie-associated protein PrP 27–30 in water by infrared spectroscopy. *Biochemistry* **30**, 7672–7680

31. Rogers, M., Yehiely, F., Scott, M. & Prusiner, S.B. (1993) Conversion of truncated and elongated prion proteins into the scrapie isoform in cultured cells. *Proc. Natl. Acad. Sci. U.S.A.* **90**, 3182–3186

32. Kellings, K., Meyer, N., Mirenda, C., Prusiner, S.B. & Riesner, D. (1992) Further analysis of nucleic acids in purified scrapie prion preparations by improved return refocusing gel electrophoresis. *J. Gen. Virol.* **73**, 1025–1029

33. Weissmann, C. (1991) A 'unified theory' of prion propagation. *Nature (London)* **352**, 679–683

34. Milner, J. & Medcalf, E.A. (1991) Cotranslation of activated mutant p53 with wild type drives the wild type p53 protein into the mutant conformation. *Cell* **65**, 765–774

10

Ribozymes

Helen A. James* and Philip C. Turner

*Department of Biochemistry, University of Liverpool, PO Box 147, Liverpool L69 3BX, U.K. and *School of Biological Sciences, University of East Anglia, Norwich, Norfolk NR4 7TJ, U.K.*

Introduction

One of the more important discoveries in the field of molecular biology in the past decade was that of Cech and Altman. They found that RNA molecules, once thought to be primarily passive carriers of genetic information, can carry out functions that had been ascribed to proteins. Some are capable of acting as enzymes — cleaving themselves or other RNA molecules. This revelation has led to the speculation that catalytic RNA molecules (ribozymes), with the combined roles of information carrier and catalyst, may have been the primordial molecules from which life evolved.

The features that allow ribozymes to bind and cleave their specific substrates are rapidly being identified. This work has led to the realization that ribozymes can be manipulated to bind and cleave novel substrates. These custom-designed RNAs have great potential as therapeutic agents and are becoming a powerful tool for molecular biologists.

This review describes RNA-mediated catalysis, the different forms it can take and how it can be manipulated.

RNA catalysis

An RNA catalyst, or ribozyme, is an RNA molecule that has the intrinsic ability to break and/or form covalent bonds (Kruger *et al.*, 1982). Like its protein counterpart, a ribozyme greatly accelerates the rate of a biochemical reaction and shows extraordinary specificity with respect to the substrates it acts upon and the products it produces.

Ribozymes can cleave either the 3'- or 5'-phosphodiester bond in an RNA molecule, resulting in 3'-OH and 5'-phosphate, or 5'-OH and 3'-phosphate or 2',3'-cyclic phosphate groups, respectively (Figure 1) (Cedergren, 1990). These reactions differ in the nucleophile that initiates the phosphodiester bond cleavage. If the nucleophile is the 2'-OH adjacent to the scissile phosphate bond, then 3'-phosphates or 2',3'-cyclic phosphates result, whereas if the nucleophile is a 3'-OH from an external source then 5'-phosphates are the result.

Normally, 3'-OH groups and phosphodiester bonds are exceedingly unreactive chemical species; however, it is thought that the folded RNA structure confers special reactivity to certain bonds by making them more prone to nucleophilic attack. The RNA structure aligns the reactive groups in close proximity to each other, in the correct orientation, and lowers the activation energy of the reaction by stabilizing the transition state.

RNA catalysis has been found in such diverse situations as group I and II introns, the genomes of viroids, virusoids and satellite RNAs of a number of

Figure 1. Possible patterns of phosphodiester bond cleavage
The reaction products from breaking the 5'-phosphodiester (5'-C-O-P) bond (1) or the 3'-phosphodiester (3'-C-O-P) (2) are shown (Cedergren, 1990).

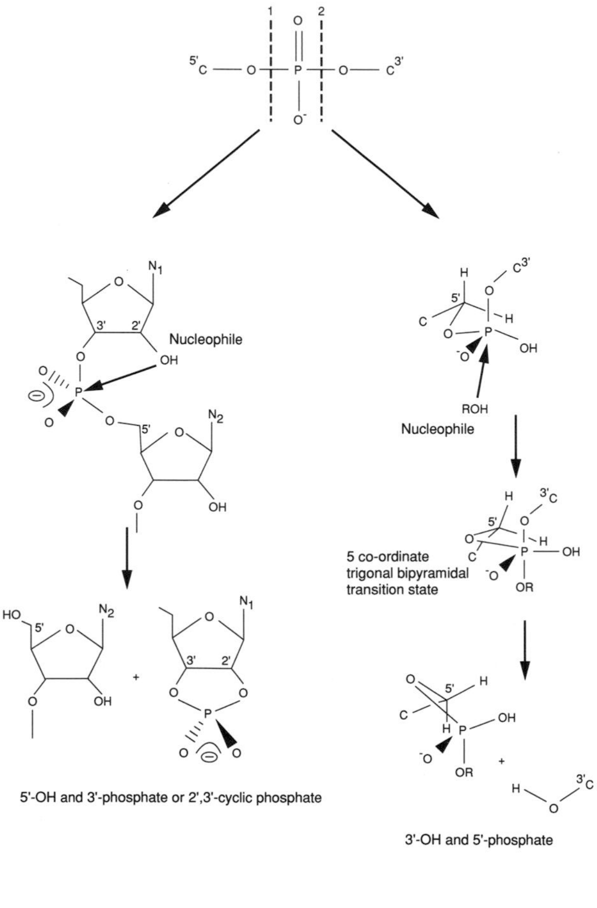

viruses, and in the prokaryotic pre-tRNA-processing machinery. Each of the catalytic RNAs will be considered in turn.

Self-splicing RNAs

Many genes have their coding sequences (exons) interrupted by stretches of non-coding DNA called intervening sequences (IVS) or introns. Transcripts of such 'split' genes, which include both exon and intron sequences (precursor-RNA) must undergo a cleavage–ligation reaction, or RNA splicing, to produce the mature functional form of the mRNA, rRNA or tRNA. Based on conserved nucleotide sequences and/or structures within and adjacent to the introns, introns have been categorized into four groups: group I, group II, nuclear mRNA and nuclear tRNA. Some examples of group I and II introns involve the RNA molecule, not only as substrate but also as a catalyst or quasi-catalyst, i.e. group I and II introns are self-splicing in the absence of protein *in vitro*.

Group I introns

Group I intron self-splicing in the absence of protein, but with guanosine, magnesium and salt, was first observed *in vitro* for the IVS of the nuclear 26S rRNA gene in *Tetrahymena thermophila* (Kruger *et al.*, 1982) and has since been observed in a number of diverse sources. These include three introns of the bacteriophage T4[1] (see also Zaug *et al.*, 1983), chloroplast *psbA* introns[2] and in the nuclear small subunit of the ribosomal RNA gene of the unicellular green alga *Ankistrodesmus stipitatus*[3].

Splicing

Group I intron splicing proceeds by two consecutive transesterification reactions (Figure 2a) (Zaug *et al.*, 1983). No net change in the number of ester linkages occurs, so no external energy source (e.g. ATP hydrolysis) is required. The first reaction is initiated by a nucleophilic attack by the 3′-OH of a guanosine or one of its phosphorylated derivatives (GMP, GDP or GTP) at the phosphodiester bond between the 5′ exon and the intron (5′ splice site). This forms a 3′,5′-phosphodiester bond between the guanosine and the first nucleotide of the intron. The new 3′-OH group of the 5′ exon then initiates a second nucleophilic attack, this time on the phosphodiester bond between the 3′ exon and the intron (the 3′ splice site). This results in ligation of the exons and excision of the intron.

The RNA cleavage and ligation activities are intrinsic to the structure of the RNA molecule. Group I introns have a common core containing a set of paired regions. The 5′ splice site is aligned for accurate splicing by base-pairing between the internal guide sequence, near the 5′ end of the intron, and the 5′ exon[4]. The 3′ splice site (always preceeded by a guanosine) is aligned by further base-pairing[5] and tertiary interactions, and the 5′ and 3′ exons are held in close proximity to the guanosine-binding site.

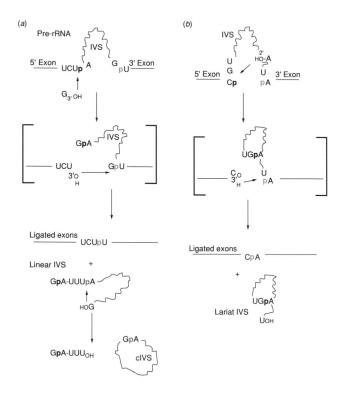

Figure 2. Self-splicing of Group I and II introns
(a) Group I introns: splicing of the *Tetrahymena* pre-rRNA IVS and cyclization of the excised IVS
by consecutive transesterifications. (b) Group II introns with the formation of a lariat: straight
lines indicate exons; wavy lines indicate IVS; [ ], transient intermediate; **p**, 5′-splice-site phosphate; p (blue), 3′-splice-site phosphate; p (grey), cyclization-site phosphate.

   After its excision from the pre-rRNA, the *Tetrahymena* intron undergoes
an intramolecular cyclization reaction (Figure 2a) (Kruger *et al.*, 1982). Again
this is a transesterification reaction and is self-catalysed. The 3′ terminal
G-OH attacks a phosphodiester bond near the 5′ end of the intron, liberating
a small linear fragment and creating a new 5′,3′-phosphodiester bond.
   Self-splicing is, by definition, an intramolecular event. The IVS greatly
speeds up the chemical reaction by lowering the activation energy; it has high
specificity but it does not emerge unaltered from the splicing reaction. It is,
therefore, not acting as a true enzyme. However, the catalytic activity found
within the conserved core can be dissociated into a distinct active enzyme core
and a distinct substrate.

The *Tetrahymena* IVS as a true enzyme
Shortened versions of the *Tetrahymena* IVS (L-19 IVS and L-21 *Sca*I IVS)
have been shown to be true enzymes *in vitro*. By removing 19–21 nucleotides
from the 5′ end of the IVS, the IVS remains linear (since the cyclization point
has been removed) and retains its ability to catalyse reactions.

The first reaction investigated in which the L-19 IVS acts as a true enzyme is an oligonucleotide disproportionation[6]. The substrate pC_5, for example, is acted upon, with a covalent enzyme–substrate intermediate, to produce pC_4, pC_6 and so on in multiple reactions. This reaction is template dependent: the substrate binds to the internal guide sequence of the IVS. Each step is an intermolecular version of the RNA self-splicing and IVS cyclization.

Zaug and colleagues showed that the L-19 IVS could act as an RNA restriction endonuclease, by a mechanism that involved guanosine transfer[7]. In this case an external guanosine is used and there is no covalent, linked intermediate with the enzyme. Again the substrate base-pairs to the internal guide sequence. If the internal guide sequence is mutated, the sequence specificity for the substrate can be altered.

Other reactions that the *Tetrahymena* IVS can catalyse include the elongation of a substrate (template dependent) by nucleotidyl transfer at the expense of dinucleotides[8], phosphotransfer[9] and cleavage of single-stranded DNA by a variant of the L-21 *Sca*I IVS, which was selected *in vitro*[10]. The latest addition to the list of catalytic activities of the *Tetrahymena* ribozyme is the discovery that it can hydrolyse the aminoacyl ester bond of *N*-formylmethionyl-tRNA[fmet 11].

Despite all the evidence for self-splicing *in vitro*, it is clear that splicing *in vivo* requires protein factors to play important roles. Even the *Tetrahymena* IVS, which at low levels of Mg^{2+} splices efficiently *in vitro*, is splicing at a rate of about 50-fold less than the level estimated for splicing *in vivo* (Kruger *et al.*, 1982).

Group II introns

Group II introns, which are classed together on the basis of conserved sequences and secondary structure, are found in organellar genes of lower eukaryotes and plants. They are excised by a two-step splicing reaction, two transesterifications, to produce branched circular RNAs, called lariats. It has been demonstrated that some members of the group, like some group I introns, can self-splice *in vitro* (Peebles *et al.*, 1986).

The unimolecular reaction *in vitro* was shown to be Mg^{2+} dependent (≥ 5 mM); to require spermidine; to have a temperature optimum of 45 °C; to have a requirement for ribonucleotides or monovalent cations as co-factors; and to require a pH of between 6.5 and 8.5 (Peebles *et al.*, 1986). The lariat that is excised is non-linear RNA with the unusual 2′,5′-phosphodiester bond at the branch point.

Splicing

Group II introns splice by way of two successive phosphate-transfer reactions (Figure 2*b*). In the first step, the 2′-OH group of an intramolecular branch-point adenosine attacks the phosphodiester bond at the 5′ splice site (creating a 2′,5′-phosphodiester bond), producing the free 5′ exon and a splicing intermediate, the intron–3′ exon. The second step involves cleavage at the 3′ splice site by the 3′-OH of the 5′ exon. Simultaneously, the exons are ligated and the intron lariat is released.

Between the two steps, the free 5′ exon is 'held' by base-pairing between the exon-binding site and the intron-binding site. This ability of the group II introns to specifically bind the 5′ exon has been exploited to encourage the IVS to catalyse reactions on exogenous substrates. Investigations have now shown that, as well as a reversal of splicing[12], a group II intron can ligate single-stranded DNA to RNA, and another can cleave a single-stranded DNA substrate. Eventually this ability to cleave DNA could be used to develop group II intron ribozymes targeted against selected sites of DNA substrates, thus allowing the application of specific restriction endonucleases *in vitro*.

Nuclear pre-mRNA splicing

The splicing of introns from nuclear pre-mRNA resembles that of group II intron splicing, but has not yet been shown to be catalysed by RNA. Like the self-splicing mechanism of group II introns, the exons of nuclear pre-mRNA are spliced in a two-step process with the production of a lariat–IVS. The introns, however, lack the secondary structures necessary to form catalytic sites and, therefore, splicing is mediated by small nuclear ribonucleoprotein particles (snRNPs) and other protein factors.

The snRNPs U1, U2, U5 and U4/U6 participate in nuclear pre-mRNA splicing (reviewed by Maniatis and Reed, 1987). The snRNPs consist of a small nuclear RNA (snRNA) moiety and several common and specific proteins. These assemble on the pre-mRNA in a concerted manner, to form the so-called spliceosome in which splicing occurs. This involves certain base-pairing interactions between the different RNA molecules. Although the snRNAs have not yet been shown to be catalytically active, there is a potential for them to show ribozyme activities in the splicing of pre-mRNA.

Ribonuclease P

Ribonuclease P (RNase P) is the ubiquitous endoribonuclease that processes the 5′ end of precursor tRNA molecules (reviewed in Darr *et al.*, 1992). It cleaves specific 3′-phosphodiester bonds to produce 5′-phosphate and 3′-OH termini and requires a divalent metal ion *in vitro*. RNase P consists of a protein moiety and an RNA moiety. It was discovered that, at least in bacteria, it is the RNA moiety that is the catalyst (Guerrier-Takada *et al.*, 1983), and as such is the only ribozyme that is a true enzyme: it catalyses an intermolecular reaction, is unaltered by the event, follows Michaelis–Menten kinetics and is stable.

The best-studied RNase P endonucleases are from *Escherichia coli* (M1 RNA and C5 protein) and *Bacillus subtilis* (P RNA and P protein). It was found that at high ionic strengths (60 mM Mg^{2+}) the RNA could cleave pre-tRNA *in vitro* in the absence of protein (Guerrier-Takada *et al.*, 1983). It is thought that *in vitro* the high ionic strengths counter the repulsion between the anionic enzyme and substrate, and that *in vivo* the protein screens the electrostatic repulsion of the RNAs, facilitating binding and thus cleavage. It is not clear what the mechanism of cleavage is, but it has been proposed that an S_N2 in-line displacement mechanism with a solvated Mg^{2+} as nucleophile may occur.

RNase P does not require a 2'-OH at the cleavage site, there is no requirement for a particular nucleotide sequence on the 5' side of the cleavage site, nor is there a need for a terminal 3'-OH group in the intact M1 RNA. What does seem to be required is the 3'-proximal CCA sequence of the pre-tRNA acceptor stem[13], and a half-turn of helix. The sequence of the acceptor stem may act as an external guide sequence to the ribozyme[13]. Exploiting this external guide sequence it may be possible to alter the RNase P specificity to cleave substrates that lack conserved features of the natural substrate, if an additional small RNA with these features (sequence complementary to the substrate and a 3'-proximal CCA sequence) is included. In principle, any RNA could be targeted by custom-designed external guide sequence RNA for specific RNase P cleavage *in vitro* or *in vivo*.

Self-cleaving RNAs

The other category of intramolecular RNA catalysis is that producing 2',3'-cyclic phosphate and 5'-OH termini on the products.

A number of small plant pathogenic RNAs (viroids, satellite RNAs and virusoids), an RNA transcript from satellite II DNA of the newt, a transcript from a *Neurospora* mitochondrial DNA plasmid and the animal virus HDV (hepatitis delta virus) undergo a self-cleavage reaction *in vitro* producing these termini in the absence of protein (Table 1). The reactions require Mg^{2+} and neutral pH[14,18].

Rolling circle replication

In the case of the pathogenic RNAs, it is thought that the self-cleavage reaction is an integral part of their replication. The circular plant pathogenic RNAs are considered to undergo replication by a rolling circle mechanism *in vivo* (Branch and Robertson, 1984), the two possible variations of which are shown in Figure 3. Circular, monomeric, genomic (plus)- and antigenomic (minus)-RNA, or concatameric minus-RNA for the second pathway, act as templates for synthesis of longer-than-unit-length precursor RNAs. The production of monomeric forms from concatamers requires highly specific cleavage. It is most likely that an RNA-catalysed, self-cleavage event performs this task. This was demonstrated *in vitro* for plus- and minus-sTRSV (satellite RNA of tobacco ringspot virus)[17] and plus- and minus-ASBV (avocado sunblotch virus)[14], and later for the other pathogenic RNAs listed in Table 1. No direct evidence has been produced as yet to show that this is what occurs *in vivo*.

Table 1. Plant and animal pathogenic RNAs and other RNAs that self-cleave *in vitro*

	Strand cleaved	Reference
RNA cleaved by hammerhead structure		
Avocado sunblotch virus (ASBV)	Plus and minus	14
Peach latent mosaic virus (PLMV)	Plus and minus	15
Potato spindle tuber virus (PSTV)	Plus and minus	16
Satellite RNA of tobacco ringspot virus (sTRSV)	Plus	17
Satellite RNA (virusoid)of Lucerne transient streak virus (vLTSV)	Plus and minus	18
Transcripts from satellite II of the newt	Plus	19
RNA cleaved by hairpin structure		
sTRSV	Minus	Hampel & Tritz, 1989; Berzal-Herranz *et al.*, 1993
RNA cleaved by other structures		
(see text for details)		
Hepatitis delta virus (HDV)	Plus and minus	20, 21, Wu *et al.*, 1989
Transcripts from *Neurospora* mitochondrial DNA plasmid	Plus	22

Self-cleavage

The reaction is a simple, non-hydrolytic cleavage whereby the inter-nucleotide bond undergoes a phosphoryl-transfer reaction in the presence of Mg^{2+} (thought to be involved in the formation of the active tertiary structure). Attack by the 2′-OH on the phosphorus leads to 2′,3′-cyclic phosphate and 5′-OH termini.

Hammerhead ribozymes

These self-cleaving RNAs can be subdivided into groups depending upon the sequence and secondary structure formed about the cleavage site. The first group of RNAs all have the same structural motif (2-dimensional), known as a 'hammerhead', which has been shown to be sufficient to direct site-specific cleavage (Figure 4a) (Uhlenbeck, 1987). The features of the hammerhead structure are three base-paired stems (helices I, II and III) which flank the susceptible phosphodiester bond, and two single-stranded regions, of which 13 nucleotides are highly conserved in sequence[18] (see also Uhlenbeck, 1987). Extensive mutagenesis (see below) has revealed sequence requirements and elucidated the important nucleotides and functional groups for efficient catalysis.

In two cases — minus-RNA of ASBV and the newt satellite transcripts — the hammerhead structure appears weak, owing to a sterically constraining

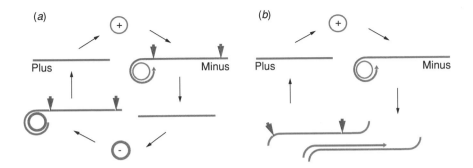

Figure 3. Rolling circle replication
(*a*) Both plus- and minus-strands can self-cleave. Circular plus-RNA is the template for the synthesis of concatameric minus-RNA. This self-cleaves to produce monomers which are then circularized. These then act as templates for the synthesis of concatameric plus-RNA. This self-cleaves and is circularized. (*b*) Only the plus-strand self-cleaves. Concatameric minus-RNA is transcribed and is used as a template for the synthesis of concatameric plus-RNA, which then self-cleaves and is circularized.

loop 2 and helix II. An alternative, double-hammerhead structure[23] (Figure 4*b*) in which two single-hammerhead domains combine has been shown to be far more stable. The minus-RNA of ASBV was shown to utilize the double-hammerhead during transcription, but the single-hammerhead structure in standard cleavage conditions, indicating the importance of the pathway of RNA folding in determining which active structure is utilized.

Hammerhead *trans*-cleavage

Uhlenbeck (1987) realized the potential for *trans*-cleavage reactions, in which two independent RNAs (based on the ASBV catalytic centre) can interact to form an active hammerhead structure. The hammerhead is split along the line B–C (Figure 5*a*). The smaller fragment with some of the conserved nucleotides, which remains unaltered and can participate in many cleavage reactions, cleaves the longer fragment that contains the GUC target site and the rest of the conserved nucleotides.

An active hammerhead structure has also been constructed from three independent RNAs (split along A–B–C)[24]. This ribozyme contains only 11 of the 13 conserved nucleotides, and indicates that not all the nucleotides conserved *in vivo* are required for the formation of active hammerheads *in vitro*.

In the above models of *trans*-cleavage, the substrate RNA contains a lot of the conserved nucleotides, and the ribozyme the rest of them. Haseloff and Gerlach (1988) proposed a model whereby the hammerhead domain is split along A–B so that the substrate RNA has only the conserved nucleotides adjacent to the cleavage site, whereas the ribozyme has all the other conserved nucleotides of the catalytic core (Figure 5*b*).

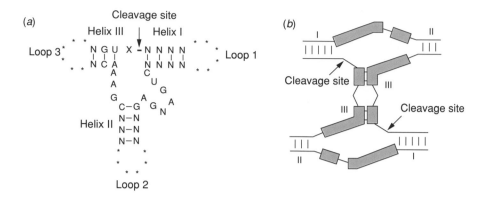

Figure 4. Hammerhead ribozyme structures

(a) Conserved sequences associated with naturally occurring cleavage sites in (plus) ASBV, sTRSV and vLTSV, for example (see Table 1 for abbreviations). N is any nucleotide; X is A, C or U; * denotes loops of RNA which can be 100–200 nucleotides in length. (b) Schematic diagram of a double-hammerhead structure, such as that associated with minus-ASBV. Conserved sequences are shaded. Helices are numbered as in (a).

Mutagenesis

The relative simplicity of the hammerhead structure and the ability to perform *trans*-cleavage reactions has made it possible to study the role of the conserved nucleotides in the catalytic core and the mechanism of cleavage. This work has revealed that the target site, commonly a GUC sequence, can be extended to NUY (where Y = A, C or U; N is any nucleotide)[25] (see also, Haseloff and Gerlach, 1988). Further investigations have revealed roles for the 2-amino and 2'-OH groups of G_5, the N^7-nitrogen of A_6 and the 2'-OH of G_8, the latter possibly in the binding of the Mg^{2+} ion[26,27]. Mutation analysis has also allowed determination of the minimum sequence required for catalytic activity. This has resulted in the 'minizyme' of McCall and colleagues[28].

Hammerhead ribozymes as endonucleases

Haseloff and Gerlach's model (Figure 5*b*) (Haseloff and Gerlach, 1988) has allowed the design of specific endoribonucleases, each with a hammerhead catalytic domain. The idea is to find NUY sequences in the substrate RNA, to define the cleavage site, and to design the flanking arms of the ribozyme (helices I and III) to bring the hammerhead catalytic core to the correct position for cleavage. Haseloff and Gerlach tested this model by designing three ribozymes against the mRNA for chloramphenicol acetyl transferase (CAT), all of which cleaved *in vitro*.

This work has encouraged others to design *trans*-acting ribozymes. The ability to cleave the RNA, and thereby inhibit the expression of a specific gene selectively, has two main applications: as a tool for molecular biology (manipulation of RNAs *in vitro*) and the inactivation of gene transcripts *in vivo*, as anti-viral agents for example.

(a)

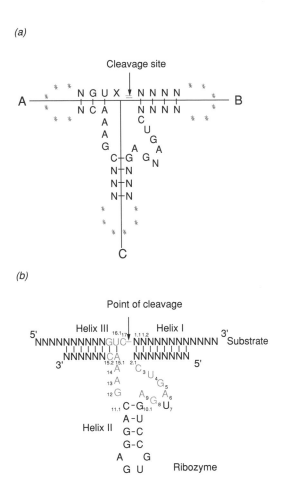

(b)

Figure 5. Hammerhead *trans*-cleavage
(*a*) The possible ways of splitting the hammerhead structure into two or more independent molecules, along the lines A, B and C. (*b*) Model for the design of hammerhead ribozymes (Haseloff and Gerlach, 1988) split along the line A–B. The catalytic domain is based on the sequence of the active site of plus-sTRSV. Conserved sequences are blue, numbering is taken from reference 52.

Hammerhead ribozymes have been used as molecular biological tools in several ways. When transcribing RNAs *in vitro* using bacteriophage polymerases it is relatively easy to define the 5′ end of the transcript (by placing the sequence to be transcribed immediately after the promoter) but it is less easy to define the 3′ end. A hammerhead ribozyme, which is part of the initial transcript, has been designed to cleave and thus produce a defined 3′ end of the transcript[29]. Ribozymes have also been used to study the roles of particular genes[30].

The targeted destruction of viral and/or cellular mRNA, to eliminate the formation of a protein that is deleterious in a disease state, has potential therapeutic value. Before this can be accomplished using ribozymes, a number of things must be considered: the choice of cleavage site within the substrate molecule; the length of the ribozyme's hybridizing arms (in the case of the hammerhead ribozymes); the stability of the ribozyme; activity within the cell and the delivery of the ribozyme to the cell (the last is discussed later).

At present, the selection of target sites is arbitrary. The accessibility of the target site due to secondary structure of the substrate RNA is a major determinant of hammerhead catalytic efficiency and secondary structure has been shown to abolish all cleavage[31]. One approach to overcome this problem is to use multi-ribozymes that target many sites within one substrate molecule[32] in the hope that at least one will cleave.

The length of the flanking arms of the ribozyme that base-pair with the substrate either side of the cleavage site must be a compromise: enough base-pairs to provide specificity but few enough to allow dissociation. It is considered that 11–15 nucleotides define a unique sequence of RNA[33]. More nucleotides in the arms to base-pair with the substrate may, at first, seem to give the ribozyme more specificity but may not be a good idea. Flanking arms that are too long may actually result in less-specific interactions, since mismatches will be tolerated, and turnover by the ribozyme will decrease because dissociation of the products from the ribozyme will be lower[33]. Another caution is that rules governing optimal ribozyme structure *in vitro* can be radically different in the cell[34]. It was found that, while good catalytic activity was exhibited *in vitro* by ribozymes with flanking arms of <15 nucleotides, inhibition in cells was only conferred by those with flanking arms of >27 nucleotides.

With the *in vivo* systems developed so far it has been difficult to show that it is ribozyme cleavage, and not just antisense effects, that is responsible for any effect observed. This is the case with ribozymes directed against the mRNA of CAT in a transient expression experiment in monkey cells[35]. Cleavage products were not observed either when ribozyme DNA and the target RNA were co-injected into oocytes[36]. However, Saxena and Ackerman have managed to show cleavage products when their ribozymes, designed to target the 28S RNA of *Xenopus*, were microinjected into oocytes[37]. Cleavage products were also detected in plant protoplasts co-transfected with ribozyme and target genes, and the reduction of activity of the target gene was shown to be due to both ribozyme cleavage and an antisense effect[38].

In a prokaryotic system, Chaut and Galibert[39] have tested the activity *in vivo* of ribozymes against β-galactosidase mRNA. Only when *cis*-acting was the ribozyme able to cleave the target. In a second study, a ribozyme capable of cleaving its substrate RNA (mRNA for *E. coli* β-glucuronidase) *in vitro* did not produce any significant reduction in β-glucuronidase activity when assayed for activity in plant protoplasts[40].

Another problem encountered is the vast excess of ribozyme (up to 1000-fold excess has been used) over target required for efficient cleavage[35] (see also Sarver *et al.*, 1990). This demonstrates the lack of multiple turnover by these ribozymes, one of the advantages it was hoped ribozymes would have over antisense oligonucleotides.

Despite these problems, several groups are currently investigating the potential of ribozymes as anti-viral agents, and anti-HIV agents in particular (Sarver *et al.*, 1990; Rossi *et al.*, 1991).

Hairpin ribozymes

Both plus- and minus-strands of sTRSV can self-cleave *in vitro*[17] (see also Hampel and Tritz, 1989). The process is thought to occur *in vivo* and to be involved in the production of monomers from concatamers during rolling circle replication. Both these RNAs cleave to give 2′,3′-cyclic phosphate and 5′-OH termini. The sequences of the catalytic site of plus-sTRSV can be arranged into the hammerhead configuration (Figure 4a)[18], but none of the sequences near the catalytic site of the minus-sTRSV ribozyme can. Instead, the catalytic RNA–substrate RNA complex forms a 2-dimensional hairpin structure with four helical domains and five loop structures (Figure 6) (Hampel and Tritz, 1989; Berzal-Herranz *et al.*, 1993). Cleavage occurs 5′ of a GUC sequence which differs from the hammerhead ribozyme of the plus-strand which cleaves 3′ of a GUC sequence. Investigations have revealed that at the target site the nucleotide at position −1 (5′ of the scissile bond) must possess a 2′-OH, and the base at position +1 must be guanosine (Berzal-Herranz *et al.*, 1993) because the 2-amino group is an essential component of the active site.

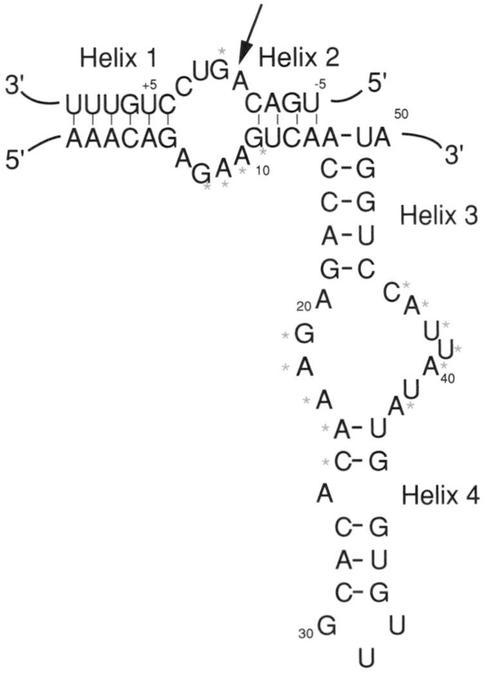

Figure 6. Secondary structure model of the hairpin ribozyme
Ribozyme sequences are numbered from 1 to 50; substrate sequences from −5 to +9. Arrow indicates the cleavage site; * denotes essential nucleotides. Modified from Berzal-Herranz *et al.* (1993)

Trans-acting hairpin ribozymes

Two helices, 1 and 2 (see Figure 6), form between the substrate and ribozyme, and it is this feature that allows the design and specificity of binding for *trans*-acting hairpin ribozymes. The hairpin ribozyme has a more complicated substrate requirement than the hammerhead ribozyme, but a substrate RNA with the sequence

$$5'\text{-NN(G/A)(U/C)N*G(A/U/C)(U/C)(G/U/C)NNNNN-}3'$$

(where N is any nucleotide and * is the bond that is cleaved) can be cleaved with high efficiency[41]. Despite this increased substrate requirement, any RNA of interest is expected to have numerous potential target sites.

The hairpin ribozyme is reported to be more catalytically efficient than the hammerhead ribozyme. The hairpin's temperature optimum is 37 °C, and it is inactive over 40 °C, probably due to the melting of secondary structure at the catalytic core (Hampel and Tritz, 1989). In this respect the hairpin is quite different from the hammerhead, since the latter's optimum temperature is 50 °C. However, both are dependent on pH and Mg^{2+}, although the hairpin has a lower Mg^{2+} concentration requirement. All this suggests that the hairpin may be better than the hammerhead *in vivo* since its optimal conditions are closer to physiological conditions.

Applications

Specifically designed *trans*-acting hairpin ribozymes have been used to cleave the *E. coli* pro-lipoprotein signal peptidase mRNA *in vitro* at three different sites[42]. Cleavage by a hairpin ribozyme of the leader sequence of human HIV-1 RNA both *in vitro* and *in vivo*, in a transient expression assay in HeLa cells, has been observed[43]. As with hammerhead ribozymes, a molar excess of ribozyme over target was required for cleavage *in vivo*, but it was demonstrated that cleavage, and not just an antisense effect, was in action.

Hepatitis delta virus

HDV plus- and minus-RNAs contain a self-cleavage site hypothesized to function in processing the viral RNA during rolling circle repliction (Wu *et al.*, 1989). Like the plant pathogens, the sites in HDV are postulated to have a related secondary structure, three models of which have been proposed: cloverleaf (Wu *et al.*, 1989), pseudoknot[20] and axehead[44], none of which is similar to the catalytic domains previously characterized (hammerhead or hairpin). It is thought that these models, each supported by computer RNA-folding predictions, nuclease-probing data and other evidence, may represent different conformations of the RNA about the cleavage site and that interconversion occurs between inactive and active forms. This adoption of alternative structures is supported by the evidence that the inclusion of denaturant (formamide or urea) can increase the cleavage activity[21].

Like the other ribozyme motifs, the HDV ribozymes require a divalent cation (Wu *et al.*, 1989), and cleavage results in products with 2',3'-cyclic phosphate and 5'-OH termini[44]. Again, the 2'-OH of the nucleotide 5' of the cleavage site is required for cleavage[21]. Only one nucleotide on the 5' side of the scissile bond[20] and ≤85 nucleotides 3' of the scissile bond[21] are necessary for efficient cleavage.

Each model has extensive stem–loop structures, the function of which is the focus of several studies[21] (see also Wu *et al.*, 1989), but so far the catalytic core of the HDV ribozyme remains obscure. Despite this, the HDV self-cleaving RNA can be separated into two independent molecules[21,44]. Investigations of *trans*-cleavage with the HDV ribozyme is only in its infancy, and, as yet, it has not been demonstrated that the HDV ribozyme can be adapted to cleave RNAs *in vivo*.

Neurospora mitochondrial VS RNA

A single-stranded circular RNA of 881 nucleotides found in the *Neurospora* mitochondrion, known as VS RNA, shares some features of the self-catalytic RNAs of HDV, group I introns and some plant viral satellite RNAs[22]. Although VS RNA can be drawn to have a secondary structure like group I introns, it is missing essential base-pairing regions, the cleavage site is in a different position and the termini produced are 5'-OH and 2',3'-cyclic phosphate. Like hammerhead ribozymes, VS RNA requires divalent cations (Mg^{2+} is best) for cleavage *in vitro* but no nucleoside. VS RNA has a temperature optimum of about 40–50 °C, is unaffected by pH over the range 5.5–8.9 and is inhibited by very low concentrations of denaturants[45].

The catalytic core of *Neurospora* VS RNA has been shown to consist of 154 nucleotides[46]. Only one nucleotide 5' of the scissile bond (like HDV) is required for cleavage. This ribozyme may represent yet another catalytic RNA group.

Ribozyme delivery

Two general mechanisms exist for introducing catalytic RNA molecules into cells: exogenous delivery of the preformed ribozyme and endogenous expression from a transcriptional unit.

Preformed ribozymes can be delivered into cells using liposomes[47,48], electroporation or microinjection. The stability of the preformed catalyst is a problem, however, as is the transient effect this approach has. Efforts have been made to increase the stability of introduced RNAs by using modified nucleotides, 2'-fluoro- and 2'-amino-[49] or 2'-O-allyl- and 2'-O-methyl-[50], mixed DNA/RNA molecules[48], or by the addition of terminal sequences (such as the bacteriophage T7 transcriptional terminator) at the 3'-end of the RNA to protect against cellular nucleases[47].

Endogenous expression has been achieved by inserting ribozyme sequences into the untranslated regions of genes transcribed by RNA polymerase II which have strong promoters, such as the SV40 early promoter[35], the actin gene (Sarver *et al.*, 1990) or a retroviral long terminal repeat[51]. A ribozyme has also been inserted into the anticodon loop of tRNA (transcribed by RNA polymerase III) and shown to be functional in *Xenopus* oocytes[36].

For this approach to work effectively the stability, intracellular localization and potential toxicity of the catalytic RNA will be critical. A number of RNA modifications, such as 5'-capping, and 3'-polyadenylation may prolong the half-life and facilitate intracellular compartmentalization of the RNAs. RNA derived from an RNA polymerase II promoter will acquire a monomethyl cap ($5'-m^7GpppG$), which may enhance its cytoplasmic stability. Most snRNA transcripts will acquire a trimethyl cap ($5'-m_3^{2,2,7}GpppG$), and, along with the binding site (for core snRNPs) found in most snRNAs, should localize the ribozyme to the nuclear compartment.

A further possible problem is adverse effects on the cells due to the toxicity of ribozymes, but so far this has not been a problem with those ribozymes studied *in vivo*.

The long-term prospects for ribozymes (hammerhead, hairpin or other types) are reasonably good despite the problems encountered: for example, the lack of specificity, target site choice and poor cleavage efficiency *in vivo* (the large excess of ribozyme over substrate required). Indeed the first human trial investigating the use of ribozymes against HIV-1 is underway in the U.S.A. (Barinaga, 1993), and initial results are promising.

Catalytic RNAs may not be limited to cleaving nucleic acids. Ribozymes have been manipulated from cleaving RNA substrates to DNA nucleases and ligases. It has also been reported that the *Tetrahymena* ribozyme can hydrolyse the aminoacyl ester bond of *N*-formylmethionyl-tRNA[fmet 11]. In the future it is possible that other ester bonds, and indeed other types of bond, may be subject to ribozyme cleavage.

Suggestions for further reading

Baringa, M. (1993) *Science* **262**, 1512–1514

Berzal-Herranz, A., Joseph, S., Chowrira, B.M., Butcher, S.E. Burke, J.M. (1993) *EMBO J.* **12**, 2567–2574

Branch, A.D. & Robertson, H.D. (1984) *Science* **223**, 450–455

Cedergren, R. (1990) *Biochem. Cell Biol.* **68**, 903–906

Darr, S.C., Brown, J.W. & Pace, N.R. (1992) *Trends Biochem. Sci.* **17**, 178–182

Guerrier-Takada, C., Gardiner, K., Marsh, T., Pace, N. & Altman, S. (1983) *Cell* **35**, 849–857

Hampel, A. & Tritz, R. (1989) *Biochemistry* **28**, 4929–4933

Haseloff, J. & Gerlach, W.L. (1988) *Nature (London)* **334**, 585–591

Kruger, K., Grabowski, P.J., Zaug, A.J., Sands, J., Gottschling, D.E. & Cech, T.R. (1982) *Cell* **31**, 147–157

Maniatis, T. & Reed, R. (1987) *Nature (London)* **325**, 673–678

Peebles, C.L., Perlman, P.S, Mecklenburg, K.L., Petrillo, M.L., Tabor, J.H., Jarrell, K.A. & Cheng, H-L. (1986) *Cell* **44**, 213–223

Rossi, J.J., Cantin, E.M., Sarver, N. & Chang P.F. (1991) *Pharm. Ther.* **50**, 245–254

Sarver, N., Cantin, E.M., Chang P.S., Zaia, J.A., Ladne, P.A., Stephens, D.A. & Rossi, J.J. (1990) *Science* **247**, 1222–1225

Uhlenbeck, O.C. (1987) *Nature (London)* **328**, 596–600

Wu, H.-N., Lin, Y.-J., Lin, F-P., Makino, S., Chang, M-F. & Lai, M.M.C. (1989) *Proc. Natl. Acad. Sci. U.S.A.* **86**, 1831–1835

Zaug, A.J., Grabowski, P.J & Cech, T.R. (1983) *Nature (London)* **301**, 578–583

This topic has not yet been comprehensively reviewed and several of the matters dealt with in the text are not referred to in the Suggestions for further reading given above. To help those wishing to go deeper into the subject, a restricted list of original papers dealing with each of the specific topics covered is given below.

References

1.   Gott, J.M., Shub, D.A. & Belfort, M. (1986) *Cell* **47**, 81–87
2.   Herrin, D.L., Bao, Y., Thompson, A.J. & Chen, Y-F. (1991) *Plant Cell* **3**, 1095–1107
3.   Davila-Aponte, J.A., Huss, V.A.R., Sogin, M.L. & Cech, T.R. (1991) *Nucleic Acids Res.* **19**, 4429–4436
4.   Garriga, G., Lambowitz, A.M., Inoue, T & Cech, T.R. (1986) *Nature (London)* **322**, 86–89
5.   Suh, E.R. & Waring, R.B. (1990) *Mol. Cell. Biol.* **10**, 2960–2965
6.   Zaug, A.J. & Cech, T.R. (1986) *Science* **231**, 470–475
7.   Zaug, A.J., Been, M.D. & Cech, T.R. (1986) *Nature (London)* **324**, 429–433
8.   Kay, P.S. & Inoue, T. (1987) *Nature (London)* **327**, 343–346
9.   Zaug, A.J. & Cech, T.R. (1986) *Biochemistry* **25**, 4478–4482
10. Robertson, D.L. & Joyce, G.F. (1990) *Nature (London)* **344**, 467–468
11. Piccirilli, J.A., McConnell, T.S., Zaug, A.J., Noller, H.F. & Cech, T.R. (1992) *Science* **256**, 1420–1424
12. Augustin, S., Muller, M.W. & Schweyen, R.J. (1990) *Nature (London)* **343**, 383–386
13. Forster, A.C. & Altman, S. (1990) *Science* **249**, 783–786
14. Hutchins, C.J., Rathjen, P.D., Forster, A.C. & Symons, R.H. (1986) *Nucleic Acids Res.* **14**, 3627–3640
15. Hernandez, C. & Flores, R. (1992) *Proc. Natl. Acad. Sci. U.S.A.* **89**, 3711–3715
16. Robertson, H.D., Rosen, D.L. & Branch, A.D. (1985) *Virology* **142**, 441–447
17. Prody, G.A., Bakos, J.T., Buzayan, J.M.M., Schneider, I.R. & Bruening, G. (1986) *Science* **231**, 1577–1580
18. Forster, A.C. & Symons, R.H. (1987) *Cell* **49**, 211–220
19. Epstein, L.M. & Gall, J.G. (1987) *Cell* **48**, 535–543
20. Perrota, A.T. & Been, M.D. (1990) *Nucl. Acids Res.* **18**, 6821–6827
21. Perrota, A.T. & Been, M.D. (1992) *Biochemistry* **31**, 16–21
22. Saville, B.J. & Collins, R.A. (1990) *Cell* **61**, 685–696

23. Epstein, L.M. & Pabon-Pena, L.M. (1991) *Nucleic Acids Res.* **19**, 1699–1705
24. Koizumi, M., Iwai, S. & Ohtsuka, E. (1988) *FEBS Lett.* **239**, 285–288
25. Perriman, R., Delces, A. & Gerlach, W.L. (1992) *Gene* **113**, 157–163
26. Fu, D-J. & McLaughlin, L.W. (1992) *Biochemistry* **31**, 10941–10949
27. Fu, D-J. & McLaughlin, L.W. (1992) *Proc. Natl. Acad. Sci. U.S.A.* **89**, 3985–3989
28. McCall, M.J., Hendry, P. & Jennings, P.A. (1992) *Proc. Natl. Acad. Sci. U.S.A.* **89**, 5710–5714
29. Grosshans, C.A. & Cech, T.R. (1991) *Nucleic Acids Res.* **19**, 3875–3880
30. Zhao, J.J. & Pick, L. (1993) *Nature (London)* **365**, 448–451
31. Xing, Z. & Whitton, J.L. (1992) *J. Virol.* **66**, 1361–1369
32. Chen, C-J., Banerjea, A.C.. Harmison, G.G., Haglund, K. & Schubert, M. (1992) *Nucleic Acids Res.* **20**, 4581–4589
33. Herschlag, D. (1991) *Proc. Natl. Acad. Sci. U.S.A.* **88**, 6921–6925
34. Crisell, P., Thompson, S. & James, W. (1993) *Nucleic Acids Res.* **21**, 5251–5256
35. Cameron, F.H. & Jennings, P.A. (1989) *Proc. Natl. Acad. Sci. U.S.A.* **86**, 9139–9143
36. Cotten, M. & Birnstiel, M.L. (1989) *EMBO J.* **8**, 3861–3866
37. Saxena, S.K. & Ackerman, E.J. (1990) *J. Biol. Chem.* **265**, 17106–17109
38. Steinecke, P., Herget, T. & Schreier P.H. (1992) *EMBO J.* **11**, 1525–1530
39. Chaut, J-C. & Galibert, F. (1989) *Biochem. Biophys. Res. Commun.* **162**, 1025–1029
40. Mazzolini, L., Axelos, M., Lescure, N. & Yot, P. (1992) *Plant Mol. Biol.* **20**, 715–731
41. Joseph, S., Berzal-Herranz, A., Chowrira, B.M., Butcher, S.E. & Burke, J.M. (1993) *Genes Dev.* **7**, 130–138
42. Tani, T., Takahashi, Y. & Ohshima, Y. (1992) *Nucleic Acids Res.* **20**, 2991–2996
43. Ojwang, J.O., Hampel, A., Looney, D.J., Wong-Staal, F. & Rappaport, J. (1992) *Proc. Natl. Acad. Sci. U.S.A.* **89**, 10802–10806
44. Branch, A.D. & Robertson, H.D. (1991) *Proc. Natl. Acad. Sci. U.S.A.* **88**, 10163–10167
45. Collins, R.A. & Olive, J.E. (1993) *Biochemistry* **32**, 2795–2799
46. Guo, H.C.T., Abreu, D.M.D., Tillier, E.R.M., Saville, B.J., Olive, J.E. & Collins, R.A. (1993) *J. Mol. Biol.* **232**, 351–361
47. Sioud, M., Natvig, J.B. & Forre, O. (1992) *J. Mol. Biol.* **223**, 831–835
48. Snyder, D.S., Wu, Y., Wang, J.L., Rossi, J.J., Swiderski, P., Kaplan, B.E. & Forman, S.J. (1993) *Blood* **82**, 600–605
49. Pieken, W.A., Olsen, D.B., Benseler, F., Aurup, H. & Eckstein, F. (1991) *Science* **253**, 314–317
50. Paolella, G., Sproat, B.S. & Lamond, A.I. (1992) *EMBO J.* **11**, 1913–1919
51. Koizumi, M., Kamiya, H. & Ohtsuka, E. (1992) *Gene* **177**, 179–184
52. Hertel, K.J., Pardi, A., Uhlenbeck, O.C., Koizumi, M., Ohtsuka, E., Uesugi, S., Cedergren, R., Eckstein, F., Gerlach, W.L., Hodgson, R. & Symons, R.H. (1992) *Nucleic Acids Res.* **20**, 3252

11

Protein stability at high temperatures

D.A. Cowan

Department of Biochemistry and Molecular Biology, University College London, Gower Street, London WC1E 6BT, U.K.

Introduction

Structural stability is an intrinsic if highly variable property of every protein. A certain level of stability is critical for the protein's ability to function successfully, either *in vivo* or *in vitro*. This stability usually reflects the environment in which the protein is synthesized and in which it must catalyse its native reaction (e.g. at 37 °C and neutral pH in a mammalian cell, or at 85 °C and pH 3 in a thermoacidophilic micro-organism). Marginal stability may be critical for functional control via rapid degradation and resynthesis, or high stability and long residence time may reflect some structural role or extracellular function.

A detailed understanding of the molecular basis of protein stability and the mechanisms of protein denaturation and unfolding encompasses many of the fundamental principles of biochemistry. The thermodynamics of protein folding and the folded state, the kinetics of folding and unfolding processes, the conformational mobility of the folded structure, the stabilizing intramolecular interactions and the interactions between the protein and the external environment (solvent and solutes) are all important factors in expanding our understanding of the behaviour of macromolecules.

Thermostable proteins may be derived from any source; however, a fundamental requirement for survival in a high-temperature environment is macromolecular stability, guaranteeing that proteins from thermophilic (i.e. high temperature) organisms will be thermostable to some degree. These organisms thus provide the ideal resource for studies on the mechanisms of protein thermostability.

Thermostable proteins and stabilization technologies are of considerable relevance to biotechnology. The successful application of enzymes in industrial processes is often critically dependent on the longevity of the catalyst, particularly where non-native conditions must be employed (i.e. the presence of detergents, extremes of pH, organic solvents and so on). The identification of thermostable enzymes, often from thermophilic sources, is an increasingly important factor in the selection of industrial catalysts, and is often a more rapid and cost-effective approach than chemical or genetic engineering of stability properties in existing biocatalysts.

One important disadvantage of the use of thermophilic enzymes is the constraint imposed by the acquisition of molecular stability. Such proteins have been 'designed' to operate most effectively at or near the growth temperature of the source organism: 37 °C for mammalian enzymes and 80–100 °C for hyperthermophilic bacteria. In consequence, the catalytic rates of hyperthermophilic enzymes at low temperatures are often very low. This, more than anything else, has stimulated the study of the molecular and conformational relationships between protein stability and catalytic function.

The fields of hyperthermophilicity and protein stability are intertwined. A number of excellent reviews emphasising different aspects of this broad field have been published. For a more detailed insight, the reader is directed to reviews on hyperthermophilic microbiology[1-6], thermostable enzymes[7-10], protein stability[11-14] and high-temperature biotechnology[15-16].

Thermophilic organisms

If you need a thermostable protein, find a thermophilic organism. The basis of this recommendation is simply that thermophiles (see Table 1) have evolved

Table 1. Growth temperature parameters for living organisms

Category	$T_{min.}$	$T_{opt.}$	$T_{max.}$
Psychrophile	<0 °C	10–16 °C	<30 °C
Mesophile	>10 °C	20–37 °C	<45 °C
Thermophile*	>30 °C	>50 °C	>60 °C
Extreme thermophile	>40 °C	>65 °C	>70 °C
Hyperthermophile	>70 °C	>90 °C	<115 °C

Organisms can be (arbitrarily) divided into categories on the basis of their growth temperature characteristics: growth minima, optima and maxima. The molecular properties which specify the upper and lower growth limits are probably complex. It is believed that membrane fluidity and protein-transport functions play a role in the lower growth limit, while protein and co-factor stability may delimit the upper temperature of life. *The term 'thermophile' is often used as a general reference for any organism growing at elevated temperatures.

the ability over time to survive, grow and reproduce at high temperatures. In consequence, evolutionary protein engineering has devised complex and subtle molecular mechanisms which result in high-temperature stability. There is excellent logic in taking advantage of this evolutionary engineering of protein stability as an alternative to the time-consuming and costly process of genetically or chemically engineering additional stability into existing mesophilic enzymes. The choice of direction will ultimately depend on many factors, not least the other essential properties of the enzyme, including the application and economic rather than biochemical criteria.

Thermophilic organisms, both prokaryotes and eukaryotes, are widely distributed in our ecosystem. Geothermal fields, fumeroles and geysers, such as are found in Yellowstone National Park in the U.S.A., in New Zealand and in Iceland, and the hydrothermal vents of the deep submarine ridges provide most of the natural, high-temperature (70–110 °C) thermal environments. Lower temperature thermal habitats (40–70 °C) are commonly found, and include composting, ore dumps and heated desert soils. Artificial thermal environments, such as domestic hot-water cylinders and industrial heated water, are also heavily colonized by thermophilic micro-organisms, including cyanobacteria, bacteria and fungi. However, all of the higher-temperature thermophiles (extreme thermophiles and hyperthermophiles) belong to either the Bacteria (see examples in Table 2) or the Archaea (Table 3). The latter phylogenetic 'domain' was identified as the result of pioneering work on the alignment of 16S rRNA sequences carried out by Carl Woese and George Fox in the late 1960s[17]. The results of their work and that of many other groups have yielded a generally accepted three-domain division of living organisms (see Figure 1) where the Archaea are thought to possess characteristics most typical of the ancestral organism or 'progenote'. The domain Archaea includes three phenotypically unusual groups of micro-organisms: the thermoacidophiles, the extreme halophiles and the methanogens.

Thermostable proteins

A number of different proteins have been isolated from various hyperthermophilic organisms (see Table 4). It is notable that the hydrolases and oxidoreductases predominate in this list, a fact which may reflect both the ease of

Table 2. Characteristics of some thermophilic bacteria

Organism	$T_{opt.}$	$T_{max.}$	$pH_{opt.}$	Growth	Metabolism
Bacillus	60–70 °C	75–80 °C	7	Aerobic	Heterotrophic
Thermus	60–75 °C	75–80 °C	7	Aerobic	Heterotrophic
Clostridium	60 °C	65 °C	6	Anaerobic	Heterotrophic
Thermoanaerobium	60 °C	65 °C	6	Anaerobic	Heterotrophic
Aquifex	90 °C	95 °C	6	Aerobic	Chemoautotrophic
Thermotoga	80 °C	90 °C	5	Anaerobic	Heterotrophic

Table 3. Characteristics of some hyperthermophilic Archaea

Organism	$T_{max.}$	$pH_{opt.}$	Oxygen status	Metabolism/energy yield
Acidianus infernus	95 °C	2.0	Aerobic or anaerobic	Obligate autotroph, either aerobic ($S^0 \to SO_4^{2-}$) or anaerobic ($S^0 \to S^{2-}$)
Archaeoglobus fulgidus	95 °C	6.5	Obligate anaerobe	Chemolithoautotrophic on H_2, CO_2 and $S_2O_3^-$; chemo-organotrophic on complex media; SO_4^{2-} reduction (S^0 not reduced)
Desulfurococcus spp.	95 °C	5.5–6.5	Obligate anaerobes	Obligate heterotrophs; anaerobic S^0 respiration; fermentation
Hyperthermus butylicus	95–106 °C	7.0	Obligate anaerobe	Fermentative; energy also obtained from $H_2 + S^0 \to H_2S$ (no S^0 respiration)
Metallosphaera sedula	80 °C	1–4.5	Aerobe	Facultative autotroph ($S^0/S^{2-} \to SO_4^{2-}$) or heterotrophy on complex media
Methanopyrus kandleri	110 °C	6.0	Obligate anaerobes	Methanogen; obligate autotroph, formate, $H_2 + CO_2 \to CH_4$
Pyrobaculum spp.	103 °C	6.0	Obligate anaerobes	Facultative autotrophs; H_2/S autotrophy or heterotrophy on complex media
Pyrococcus spp.	103 °C	6.5–7.5	Obligate anaerobes	Obligate heterotrophs; S^0 respiration; fermentation
Pyrodictium spp.	110 °C	5.5	Obligate anaerobes	Obligate H_2/S^0 autotrophy
Staphylothermus marinus	98 °C	6.5	Obligate anaerobe	Obligate heterotroph with S^0 respiration, fermentation
Sulfolobus spp.	75–87 °C	3.0	Aerobic or micro-aerophilic	Facultative autotrophs; $S^0 \to SO_4^{2-}$, $Fe[II] \to Fe[III]$, $S^{2-} \to S^0$
Thermococcus celer	93 °C	5.8	Obligate anaerobe	Obligate heterotroph; S^0 respiration of complex media, fermentation.
Thermodiscus maritimus	98 °C	5.5	Obligate anaerobe	Facultative autotroph; S^0 respiration
Thermoproteus tenax	97 °C	5.5	Obligate anaerobe	Facultative autotroph; either H_2/S^0 autotrophy or fermentation of complex media

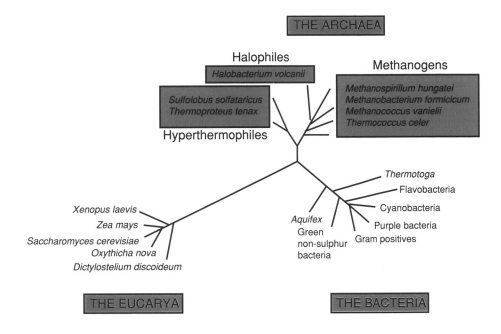

Figure 1. The 'unrooted' phylogenetic tree of life
Redrawn from Woese *et al.* (1990) *Proc. Natl. Acad. Sci. U.S.A.* **87**, 4576. The branch lengths indicate the degree of phylogenetic relatedness. The depth of the branch point indicates the proximity to the ancestral 'progenote'. As an unrooted tree, the position of the progenote is not shown.

screening for these activities and their potential application in biotechnology industries. Emphasis on other groups of enzymes (e.g. the DNA and RNA polymerases) reflects different factors: in the two examples cited, their use in polymerase chain reaction (PCR) technology and in phylogenetic studies, respectively.

These proteins have formed the basis of detailed studies on the molecular mechanisms responsible for thermostability and the relationships between protein stability, protein flexibility and catalytic activity (see below). Structurally, these proteins differ little from their mesophilic homologues. Such differences as have been identified can often be linked to their stability requirements. Extracellular enzymes are often stable at temperatures considerably higher than the growth temperature (e.g. the *Pyrococcus* α-amylases are functional at 130 °C), whereas intracellular enzymes frequently denature at temperatures very near the growth optimum, reflecting the turnover and control requirements of metabolically important enzymes. Of course, where one critical metabolic enzyme is unstable above a certain temperature, there will be no pressure on associated enzymes to evolve higher levels of stability.

A number of genes for hyperthermophilic proteins have been cloned and expressed in mesophilic hosts, including *Escherichia coli* and yeast (see Table 4). Notwithstanding the existence of protein-splicing introns in some archaeal

Table 4. Enzymes isolated from hyperthermophilic organisms

Enzyme category and type	Source	$T_{opt.}^{*}$	Comments
Hydrolases			
Protease	Pyrococcus furiosus, P. woesei, Desulfurococcus mucosus, Thermococcus celer, T. stetterii, T. littoralis, Thermococcus ANI, Staphylothermus marinus, Sulfolobus acidocaldarius Fervidobacterium pennavorans	80–115 °C	All serine proteases except S. acidocaldarius, an acid protease. This enzyme has been cloned and expressed in E. coli
Aminopeptidase	Sulfolobus solfataricus	<70 °C	
Carboxylesterase	S. solfataricus	<100 °C	Cloned and expressed in E. coli
α-Amylase	P. furiosus, P. woesei	100 °C	
Pullulanase	P. furiosus, P. woesei	105 °C	
α-Glucosidase	P. furiosus, P. woesei	110–115 °C	
β-Glycosidase	T. celer, Thermotoga Fj3, Thermococcus ANI, Desulfurococcus sp., S. solfataricus	–	The S. solfataricus β-gal has been cloned and expressed in E. coli and yeast
Cellobiohydrolase	Thermotoga Fj3	100–105 °C	
Transglycosylase	S. solfataricus, Thermotoga sp., Thermococcus ANI, Desulfurococcus sp.	90–95 °C	
Fructose 1,6-bisphosphatase	F. thermophilum	>80 °C	
Restriction endonuclease	Sulfolobus sp.	80–85 °C	SuaI
DNA topoisomerase	Methanopyrus kandleri, Desulfococcus amylolyticus, S. acidocaldarius		
ATPase	S. acidocaldarius	>75 °C	
	P. occultum	100 °C	
Xylanase	Thermotoga sp.	90–105 °C	
Oxidoreductases			
Glyceraldehyde-3-P dehydrogenase	P. woesei, T. maritima, T. tenax, Methanothermus fervidus	>90 °C†	The P. woesei GAPDH has been cloned and expressed in E. coli
Glutamate dehydrogenase	P. furiosus, P. woesei, S. solfataricus	>85 °C	The P. woesei GDH has been cloned and expressed in E. coli

Enzyme category and type	Source	$T_{opt.}$*	Comments
Oxidoreductases			
Malate dehydrogenase	S. acidocaldarius, M. fervidus	<90 °C	
Lactate dehydrogenase	S. solfataricus	>75 °C	
Malic enzyme	S. solfataricus	>85 °C	
sec-Alcohol dehydrogenase	S. solfataricus, Hyperthermus butylicus	>80 °C	The S. solfataricus enzyme has been cloned and expressed in E. coli
Aldehyde-fd-oxidoreductase	P. furiosus	>90 °C	
Pyruvate-fd-oxidoreductase	H. butylicus, T. celer	>80 °C	
Sulphur oxygenase reductase	Desulfurolobus ambivalens	85 °C	
Hydrogenase	P. furiosus, T. maritima, Methanococcus jannaschii	>95 °C	
Transferases			
DNA polymerase	P. furiosus, T. littoralis, T. maritima, S. acidocaldarius	>95 °C	P. furiosus and T. littoralis enzymes cloned and expressed in E. coli
DNA-dependent RNA polymerase	Virtually all hyperthermophiles	75–95 °C	
Propylamine transferase	S. solfataricus	90 °C	
Aspartate aminotransferase	S. solfataricus	95 °C	
Phosphofructokinase	F. thermophilum	>90 °C	
Lyases, ligases and isomerases			
DNA ligase	P. furiosus	>100 °C	Cloned and expressed in E. coli
Acetyl-CoA synthetase	P. furiosus, P. woesei, D. amylolyticus	>80 °C	
Glucose isomerase	T. maritima	-	
Citrate synthase	S. acidocaldarius, S. solfataricus	~90 °C	
Fumarate hydratase	S. solfataricus	~90 °C	

Data derived from Adams[7], Coolbear[10], Kates[4] and other sources. * The 'temperature optimum' is influenced by both Arrhenius behaviour and by thermal denaturation. The value is thus dependent on reaction conditions. The temperature optimum is usually some degrees below the denaturation temperature (T_m), which is best determined by physical methods such as microcalorimetry or spectroscopy. † Not available due to instability of the substrates at high temperatures.

DNA polymerase genes[18], no particular hurdles have been encountered in successfully expressing hyperthermophilic protein genes in mesophiles. Initial fears that newly synthesized hyperthermophilic proteins would not fold successfully in mesophilic hosts (some 50 °C below the native folding temperature) have proved unfounded.

Mechanisms of protein thermostability

The molecular mechanisms responsible for enhanced protein thermostability and the structural and functional consequences of protein thermostability have been the major focuses of thermophile research (see review by Jaenicke[12]). At biologically relevant temperatures (up to 110 °C), most covalent bonds in proteins are stable. Thus the thermodynamic stability of a protein is a balance between large stabilizing forces (derived from non-covalent intramolecular interactions) and large destabilizing forces, primarily chain conformational entropy. The total conformational energy of a small, monomeric, globular protein will be in the order of several thousands of kilojoules per mole, whereas the free energy difference between the folded and unfolded states is typically only -30 to -60 kJ·mol^{-1} [12]. Clearly, the stability of the folded state can be perturbed by relatively minor changes in primary structure. The ability of these apparently minor structural alterations to impart relatively large changes in protein stability is better understood when it is considered that typical intramolecular interactions (hydrogen bonds, salt bridges and hydrophobic interactions) can contribute between -2 and -20 kJ·mol^{-1}, depending on location and orientation.

The evolutionary consequence is a multitude of subtle molecular strategies contributing to the overall thermodynamic stability of thermostable proteins. Some of these mechanisms are summarized in Table 5.

This list is constantly being expanded by evidence from detailed structural studies, particularly those providing comparisons between closely homologous mesophilic and thermophilic proteins, supported by protein engineering studies where point mutations can be introduced to enhance, or reduce, protein thermostability (for example, see Table 3 in reference 12).

Table 5. Thermostabilization mechanisms employed in hyperthermophilic proteins

- Increase in intramolecular packing
- Loss of surface loops
- Increase in helix-forming amino acids
- Stabilization of α-helix dipoles
- Insertion of proline residues
- Reduction in asparagine content
- Restriction of *N*-terminus mobility

Inactivation at high temperatures

It is theoretically possible to introduce into proteins stabilizing interactions which could lead to stability at very high temperatures (e.g. 150 °C–200 °C). However, the constraints imposed by chemical stability define the upper limit of protein thermostabilization. Klibanov and co-workers[19] have investigated the mechanisms of irreversible thermal enzyme inactivation (Table 6). The amide side-chain of asparagine is a 'weak link', since it is particularly susceptible to deamidation at high temperatures and neutral pH. The sudden appearance of a negatively charged group in the close-packed protein interior is detrimental to the retention of a native structure. Several studies of homologous enzymes — across mesophiles, thermophiles and hyperthermophiles — have shown a progressive replacement of asparagine residues in non-catalytic positions.

The charged side-chain of aspartate can potentially attack the adjacent peptide linkage on the C-terminal side, cleaving the bond via a cyclic imide intermediate. This process has been shown to be relatively insignificant in fully folded structures, but increasingly prevalent in marginally stable (i.e. partially unfolded) proteins at elevated temperatures. Reduction of disulphide bridges and the formation of incorrectly folded disulphide interchange products may be less significant in hyperthermophiles, where many proteins lack cystine residues.

Consequences of hyperstability

The intrinsic property of thermostability (or the imposition of thermostability by chemical or genetic engineering) may impart both structural and functional constraints to a protein. It is generally believed that increasing the thermal stability of a protein results in reduced conformational flexibility. Depending on the locality and extent of changes in conformational flexibility, this may result in significant (and possibly detrimental) consequences with respect to biological function. There is now evidence from studies of proteolysis and ^{1}H–^{2}H exchange, and from calculations of flexibility profiles[20], that at any temperature a thermostable protein will be less flexible than its mesophilic homologue.

Table 6. Mechanisms of irreversible inactivation of enzymes

| | Rate constant (h^{-1}) at 100 °C | | |
	pH 4	pH 6	pH 8
Deamidation of asparagine	0.45	4.1	18
Cleavage of Asp–Xaa peptide bonds	0.12	0	0
Reduction of disulphide bonds	0	0	6
Formation of disulphide interchange structures	0	0	32

Data taken from Klibanov[19].

It is also now believed that at their respective growth temperatures, similar proteins from both mesophilic and thermophilic sources will possess similar levels of molecular flexibility, a consequence of the fact that molecular flexibility is a critical requisite for function.

It is commonly believed that thermophilicity is accompanied by the benefits of correspondingly higher catalytic rates (Figure 2*a*). While this would be perfectly reasonable if the Arrhenius temperature–kinetic relationship was the only factor governing the action of enzymes, the functional constraints imposed by the structural requirements of a stable 3-dimensional macromolecule obviously play a major role. The reality (Figure 2*b*) is that functionally similar proteins from micro-organisms of different thermal biotopes tend to possess similar catalytic rates at their respective growth temperatures. The only notable exceptions are those where the substrate is rendered more accessible/susceptible to catalytic attack at the higher reaction temperature. Hydrolases, such as amylases and proteases, are typically much more effective at higher temperatures, a fact which has been exploited for decades by the starch industry.

It has been noted on many occasions that thermostable proteins tend to have a high level of resistance to denaturation by such agents as organic solvents, detergents and chaotropic compounds, such as urea and guanidine hydrochloride. Thermostable proteins are also more resistant to degradation by proteolysis, a factor which should have practical advantages in the purification of both native and recombinant thermophilic enzymes. While the mechanisms of protein unfolding by organic solvents, detergents and so on are poorly understood, the obvious fact that thermostable proteins show enhanced resistance to all these agents implies that some aspect of the unfolding pathway is common to all (see Figure 3). It seems obvious that the first step of protein unfolding must be the rapid, reversible, conformational transition which is a

Figure 2. Perceptions and realities of the temperature–activity relationships for mesophilic and thermophilic enzymes
(*a*) The expectation; (*b*) the reality.

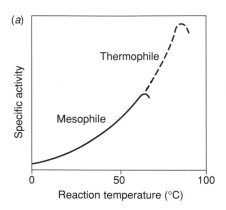

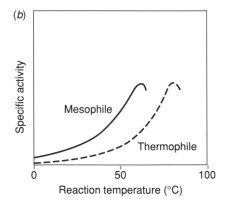

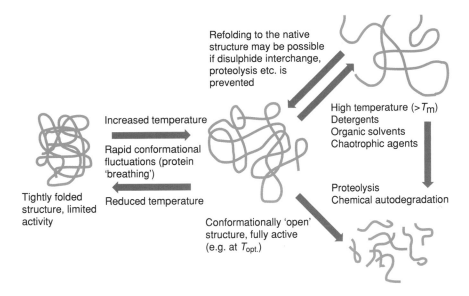

Increased temperature

Rapid conformational fluctuations (protein 'breathing')

Reduced temperature

Tightly folded structure, limited activity

Refolding to the native structure may be possible if disulphide interchange, proteolysis etc. is prevented

High temperature ($>T_m$)
Detergents
Organic solvents
Chaotrophic agents

Proteolysis
Chemical autodegradation

Conformationally 'open' structure, fully active (e.g. at $T_{opt.}$)

Figure 3. A schematic view of the processes of reversible and irreversible inactivation of enzymes

reflection of the normal flexibility of the protein. Loss of tertiary structure via denaturation (by temperature, detergent, solvent and so on) or proteolysis will only progress from the unfolded conformer. Restriction of the initial reversible conformational transitions, a consequence of thermostabilization, will proportionally reduce the tendency of the protein to partake in further irreversible unfolding steps. The resistance of thermostable proteins to proteolysis supports this scheme (Figure 3), since it is well-known that unfolded proteins show increased susceptibility to proteolytic attack.

There has been much speculation about the consequences of protein thermostability in respect of substrate specificity. On one hand, it has been suggested that the enhanced rigidity of thermostable protein structures might enhance substrate selectivity. On the other, it has been proposed that the kinetic consequences of high temperature might make active sites more fluid, thus reducing substrate specificity. Both these views ignore the now-accepted view that variations in the flexibility of the polypeptide in the region of the active site are probably highly constrained by the requirements of catalysis. Nevertheless, some enzymes from the hyperthermophilic organisms do appear to demonstrate broader specificity than their mesophilic counterparts. The examples of dual co-factor specificity, displayed by glyceraldehyde-3-phosphate dehydrogenase from *Pyrococcus woesi*, glucose dehydrogenases from *Thermoplasma acidophilum* and *Sulfolobus solfataricus*, and malate dehydrogenase from *T. acidophilum*, may indicate some relaxation of specificity, but it must be recorded that this has also been observed in a limited number of mesophilic Bacteria and Archaea. Other hyperthermophilic enzymes seem to possess a broader substrate specificity than is typical for the mesophilic equivalents (e.g. *S. solfataricus* alcohol dehydrogenase and *Sulfolobus acidocaldarius* glycosyltransferase) but there is insufficient data available at present to draw any serious conclusions.

Protein engineering for thermostability

Not surprisingly, most of the studies aimed at enhancing the stability of proteins have been carried out using mesophilic proteins. Much of the work in this area has come from the University of Oregon laboratory of Brian Matthews, whose results from the site-directed mutagenesis of lysozyme have given direct evidence for the role of individual intramolecular interactions[11]. The targeted addition of single hydrogen bonds or charge–charge interactions can frequently increase the protein melting temperature (T_m) by several degrees Celsius via thermodynamic stabilization of the folded structure.

The addition of disulphide bonds, possibly the most obvious target for stabilizing proteins, has been shown to significantly increase stability; a triple disulphide variant of lysozyme unfolded with a T_m approx. 23 °C higher than the wild-type enzyme[21]. In an analogous manner, judiciously placed Xaa → Pro and Gly → Ala replacements can also enhance structural stability[22]. Both these types of modification stabilize the folded structure by reducing the conformational entropy of the unfolded state (i.e. destabilizing the unfolded protein).

It is clearly critical that the positioning of these linkages must be designed so as to avoid any conformational constraint on the active site. Even the most carefully designed site-directed modifications can have unexpected effects on the activity of the enzyme and reduced stability is a more common consequence of site-directed mutagenesis than increased stability.

Biotechnological implications

Biotechnological applications of thermophilic enzymes are likely to reflect the single most significant property of these enzymes: the ability to function at high temperatures. This is reflected in the major commercial success of hyperthermophilic enzymes, PCR. The central core of this well-publicized and widely used technique is the enzyme DNA polymerase, providing (among other things) a rapid and efficient exponential amplification of specific sequences of DNA. The automation of PCR by replacing Klenow fragment on *E. coli* DNA polymerase with the thermostable *Taq* polymerase (from the thermophilic bacterium *Thermus aquaticus*) was a major development in the technology. However, the thermostability of *Taq* polymerase is only marginally sufficient under operational conditions and the commercial availability of extremely thermostable DNA polymerases from two different hyperthermophilic Archaea (*Thermococcus littoralis* and *Pyrococcus furiosus*) has been a further advance in the technology. The manipulation of reaction conditions (particularly salt concentrations) to enable these enzymes to be used as reverse transcriptases, thus enabling the amplification of DNA directly from mRNA, has been another critical step in the development of recombinant DNA technology.

A high degree of thermostability is not the only necessity for successful application of DNA polymerases in PCR. Other factors, such as processivity (the length of extension) and proof-reading (3',5'-exonuclease) activity are also important. For example, the proof-reading capacity of the *P. furiosus* DNA polymerase is reported to reduce the error frequency by a factor of 14, compared with *Taq* polymerase. As the uses and applications of thermophilic DNA polymerases expand[23], the development targets are inclined to change. Stimulated by the development of automated sequencing methods and the needs of the Human Genome project, the search is currently on for thermostable DNA polymerases capable of megabase chain extension.

Thermostable DNA ligases have recently attracted commercial attention because of the development[24] of a ligase-dependent, DNA-amplification technique (the ligase chain reaction; LCR or LAR). This technique can be used for the identification of single nucleotide base lesions and has considerable potential as a screening assay for the gene defects which result in serious genetic disorders, such as cystic fibrosis.

Future applications

The field of high-temperature enzymology is relatively untapped. The discovery of totally novel hyperthermophilic species or genera is currently occurring at a rate little slower than the appearance of publications on novel enzymes. Each of these new isolates represents a tremendous resource of catalytic potential. Not only can it be assumed that the hyperthermophilic Archaea and Bacteria will contain as wide a range of different enzyme activities as any other diverse group of micro-organisms, but certain activities and enzyme systems will be unique to these organisms, particularly the Archaea.

Despite optimistic predictions in the past, it seems unlikely that hyperthermophilic enzymes will 'revolutionize' industrial enzymology. In many cases, industrial biocatalytic processes are incompatible with high-temperature operation (e.g. cheese-making, beer clarification and domestic detergents) or there is no obvious process advantage in increasing the reaction temperature (e.g. biosynthetic production of fine chemicals, such as amino acids and penicillins). In other cases, enzymes of sufficient thermostability are already in use (e.g. saccharification of starch). Most importantly, many of the large industrial biocatalyst users are committed to existing low-temperature operation through investment in plant and equipment. In these cases, the advantages of higher temperature operation would be insufficient to offset the costs of replacing that equipment.

This assessment leads us to the conclusion that the commercial future of thermophilic enzymes may not be as replacements for existing 'industrial' enzymes, particularly in the high-volume, low-cost enzyme market. While this may exclude a large portion of the industrial market, the residual fraction comprises a multitude of more minor applications and other major applications loom in the near future. In some of these 'niche' applications, thermophilic enzymes will undoubtedly play a role, particularly in those instances where the unique properties of these enzymes will confer significant process advantages.

Summary

- *The enzymology of hyperthermophilic micro-organisms is a growing field. As increasing numbers of novel high-temperature organisms are isolated and made available through culture collections, and, as biomass becomes more readily available, more laboratories will undoubtedly expand their research interests into this area. The prospect of totally novel enzyme systems and of new approaches to the investigation of fundamental molecular properties will continue to stimulate interest in this field.*

- *Studies of thermostable enzymes have already provided valuable data on the relationships between protein stability and activity. The subtle molecular mechanisms which have evolved to stabilize these proteins provide the clues needed for the intelligent design of stabilized mesophilic enzymes, an important target where a combination of high activity at 'low' temperatures and resistance to denaturation is required.*

- *The current role of hyperthermophilic enzymes in biotechnology is relatively minor, despite these enzymes having a high 'profile'. While early over-enthusiastic predictions that these enzymes would revolutionize biotechnology should be disregarded, it can reasonably be assumed that, where functional and economic criteria are suitable, thermophilic enzymes will be readily incorporated into current and future biotechnology.*

References

1.   Stetter, K.O. (1986) In *Thermophiles: General, Molecular and Applied Microbiology* (Brock, T.D., ed.), pp. 39–74, Wiley, New York
2.   Cowan, D.A. (1992) Biochemistry and molecular biology of extremely thermophilic archaeobacteria, in *Molecular Biology and Biotechnology of Extremophiles* (Herbert, R.A. & Sharp, R.S., eds.), pp. 1–43, Blackie, Glasgow
3.   Kristjannson, J. (1992) *Thermophilic bacteria.* CRC Press, Boca Raton
4.   Kates, M., Kushner, D.J. & Matheson, A.T. (1993) *Biochemistry of Archaea (Archaebacteria)*, Elsevier, Cambridge
5.   Kandler, O. & Zillig, W. (1986) *Archaebacteria 85* Gustav Fischer, Stuttgart
6.   Woese, C.R. & Wolfe, R.S. (1985) Archaebacteria. *The Bacteria* **8**, 1–581

7. Adams, M.W.W. (1993) Enzymes and proteins from organisms that grow near and above 100°C. *Annu. Rev. Microbiol.* **47**, 627–658

8. Kelly, R.M., Brown, S.H., Blumenthals, I.I. & Adams, M.W.W. (1992) Characterisation of enzymes from high temperature bacteria. *ACS Symp. Ser.* **498**, 23–41

9. Kelly, R.M. & Brown, S.H. (1993) Enzymes from high temperature microorganisms. *Curr. Opin. Biol.* **4**, 188–192

10. Coolbear, T., Daniel, R.M. & Morgan, H.W. (1992) The enzymes from extreme thermophiles: bacterial sources, thermostabilities and industrial relevance. *Adv. Biochem. Eng.* **45**, 57–98

11. Matthews, B.W. (1993) Structural and genetic analysis of protein stability. *Annu. Rev. Biochem.* **62**, 139–160

12. Jaenicke, R. (1991) Protein stability and molecular adaptation to extreme conditions. *Eur. J. Biochem.* **202**, 715–728

13. Privalov, P.L. (1979) Stability of proteins. *Adv. Protein Chem.* **33**, 167–241

14. Klibanov, A.M. (1983) Stabilisation of enzymes against thermal inactivation. *Adv. Appl. Microbiol.* **29**, 1–28

15. Cowan, D.A. (1992) Enzymes from thermophilic archaebacteria: current and future applications in biotechnology. *Biochem. Soc. Symp.* **58**, 149–169

16. Bergquist, P.L. & Morgan, H.W. (1993) The molecular genetics and biotechnological application of enzymes from extremely thermophilic eubacteria, in *Molecular Biology and Biotechnology of Extremophiles* (Herbert, R.A. & Sharp, R.S., eds.), pp. 44–75, Blackie, Glasgow

17. Woese, C.R. & Fox, G.E. (1977) Phylogenetic structure of the prokaryotic domain: the primary kingdoms. *Proc. Natl. Acad. Sci. U.S.A.* **74**, 5088–5090

18. Perler, F.B., Jack, W.E., Hodges, R.A., Comb, D.G., Xu, M., Noren, C.J. & Jannasch, H. (1993) Protein splicing intervening sequences in archaeal DNA polymerase genes. *Abstr. Pap. Am. Chem. Soc.* **205**, 11

19. Ahearn, T.J. & Klibanov, A.M. (1985) The mechanism of irreversible enzyme inactivation at 100°C. *Science* **228**, 1280–1284

20. Vihinen, M. (1987) Relationship of protein flexibility to thermostability. *Protein Eng.* **1**, 477–480

21. Matsumara, M., Signor, G. & Matthews, B.W. (1989) Substantial increase of protein stability by multiple disulphide bonds. *Nature (London)* **342**, 291–293

22. Matthews, B.W., Nicholson, H. & Becktel, W.J. (1987) Enhanced protein thermostability from site-directed mutations that decrease the entropy of unfolding. *Proc. Natl. Acad. Sci. U.S.A.* **84**, 6663–6667

23. Erlich, H.A., Gelfand, D. & Snitsky, J.J. (1991) Recent advances in the polymerase chain reaction. *Science* **252**, 1647–1651

24. Barany, F. (1991) Genetic-disease detection and DNA amplification using cloned thermostable ligase. *Proc. Natl. Acad. Sci. U.S.A.* **88**, 189–193

Subject index